D0871948

The Remote Sensing Data Book

The Remote Sensing Data Book provides a unique resource of all aspects of remote sensing for the expert and non-expert alike. Organised as a dictionary, it contains over 700 alphabetically-arranged and cross-referenced entries on how remote sensing works, what kinds of data are available, and the large number of satellites and instruments from which the information is obtained. As well as short technical definitions, it also includes longer essays and reviews to give an overview of the subject. Although not a textbook in itself, the data book will serve as a valuable addition to existing textbooks for undergraduates and graduate students taking geography, environmental and earth science courses that include an element of remote sensing. It will also be an essential reference for researchers and research managers at all levels who use spaceborne remote sensing methods to obtain information about the Earth's land, sea, ice and atmosphere.

GARETH REES read natural sciences at the University of Cambridge, specialising in physics and theoretical physics, and took a PhD in radio astronomy. He is a fellow of Christ's College, Cambridge, and Head of the Remote Sensing Group at the Scott Polar Research Institute. His research interests relate to the application of satellite remote sensing methods to polar environments, especially in the Arctic. Currently his main research looks at environmental damage in the north of Russia, including air pollution and oil spills. He has published several books, including *Physical Principles of Remote Sensing* (1990), *Physics by Example* (1994), and *Essential Quantum Physics* (1997), all with Cambridge University Press.

The Remote Sensing Data Book

GARETH REES

CAMBRIDGE
UNIVERSITY PRESS

PUBLISHED BY THE PRESS SYNDICATE OF THE UNIVERSITY OF CAMBRIDGE
The Pitt Building, Trumpington Street, Cambridge, United Kingdom

CAMBRIDGE UNIVERSITY PRESS
The Edinburgh Building, Cambridge CB2 2RU, UK http://www.cup.cam.ac.uk
40 West 20th Street, New York, NY 10011-4211, USA http://www.cup.org
10 Stamford Road, Oakleigh, Melbourne 3166, Australia

First published 1999

Printed in the United Kingdom at the University Press, Cambridge

Typeset in 9/12pt Times [wv]

A catalogue record for this book is available from the British Library

Library of Congress Cataloguing in Publication data

Rees, Gareth (William Gareth), 1959–
The remote sensing data book / Gareth Rees.
  p.   cm.
ISBN 0 521 48040 X
1. Remote sensing – Handbooks, manuals, etc. 2. Remote sensing –
Dictionaries. I. Title.
G70.4.R435 1999
621.36'78–dc21   98-30283 CIP

ISBN 0 521 48040 X hardback

For Christine

Contributors

Adigun A. Abiodun, United Nations Office for Outer Space Affairs, Vienna, Austria

Phillip A. Arkin, NOAA National Center for Environmental Prediction, Washington, DC 20233-9910, USA

Robert M. Atlas, NASA Goddard Space Flight Center, Greenbelt, MD 20771, USA

Arthur P. Cracknell, University of Dundee, UK

Paul J. Curran, Department of Geography, University of Southampton, UK

Donna Demac, Communications Attorney, New York, USA

James Foster, NASA Goddard Space Flight Center, Greenbelt, MD 20771, USA

Adrian K. Fung, University of Texas at Arlington Wave Scattering Research Center, USA

James R. Heirtzler, NASA Goddard Space Flight Center, Greenbelt, MD 20771, USA

Kendal McGuffie, University of Technology, Sydney, Australia

Peter J. Mouginis-Mark, Hawaii Institute of Geophysics and Planetology, University of Hawaii, USA

W. Gareth Rees, Scott Polar Research Institute, University of Cambridge, UK

Irwin Scollar, Universität Köln, Germany

John Turner, British Antarctic Survey, UK

Fawwaz T. Ulaby, Electrical Engineering and Computer Science, University of Michigan, USA

Preface

What is remote sensing?

Remote sensing is, rather broadly speaking, the obtaining of information about an object without actually coming into contact with it. A more restricted definition includes the stipulations that the object is located on or near the Earth's surface, that the observations are made from above the object, and that the information is carried by electromagnetic radiation, some measurable property of which is affected by the object being sensed. This narrower definition excludes such techniques as sonar, geomagnetic and seismic sounding, as well as medical imaging, but includes a wide and fairly coherent set of techniques, often known by the alternative name of *Earth Observation*, that form the subject of this book.

Remote sensing can be viewed as an extension of aerial photography, and thus dated to (probably) 1858, when Tournachon made the first aerial photograph from a balloon at a height of about 80 metres. In the century and a half since then, three principal developments can be identified: the possibility of using aircraft (from the early years of the twentieth century) and spacecraft (from the 1960s) as platforms to carry sensors; the development of sensors exploiting a much wider range of the electromagnetic spectrum than is responded to by photographic film; and the computer revolution, which can again probably be dated to the 1960s and which still appears to be in full swing.

Remote sensing has seen dramatic growth over the last few decades. In part, this can be attributed to the technical developments just outlined, but it is clear that remotely sensed data (in the sense in which I have defined them) must also have some tangible advantages to justify the expense of acquiring and analysing them. These advantages derive from a number of characteristics of remote sensing. Probably the most important of these is the fact that data can be gathered from a large area of the Earth's surface (or a large volume of the atmosphere) in a short space of time, allowing a virtually instantaneous 'snapshot' to be obtained. For example, the Landsat Thematic Mapper, a spaceborne instrument, can acquire data from an area 185 km square in about half a minute. When this aspect is combined with the fact that airborne and spaceborne systems can obtain information from locations that would be difficult (slow, expensive, dangerous, politically awkward) to measure *in situ*, the potential power of remote sensing becomes apparent. Further advantages derive from the fact that most remote sensing systems now generate calibrated digital data which can be manipulated in a computer.

Remote sensing finds a very wide range of applications including, famously, military reconnaissance. The great majority of the non-military applications, with which this book deals, can be loosely categorised as 'environmental', and we can distinguish a range of environmental properties that can be sensed. In the atmosphere, these include temperature, humidity, precipitation, and the spatial distribution of clouds, winds, aerosols and minor constituents, especially ozone. The Earth's radiation budget can also be measured. Over land surfaces, topography, temperature, albedo, vegetation type and distribution, rock type, soil moisture and land use can all be measured. Over ocean surfaces, colour (which is often related to biological productivity), temperature, topography (from which surface currents can be inferred), wind velocity and wave spectra can be measured. Finally one should mention the cryosphere, where the distribution and condition of snow, sea ice, icebergs and glaciers and ice sheets can be monitored.

This wide range of measurable physical properties generates a correspondingly wide range of applications. Again it is not practicable to present an exhaustive list, but the major non-military applications can perhaps be categorised as follows:

Atmosphere
> weather forecasting
> stratospheric chemistry
> global climate research

Land surface
> cartography
> land-use surveying
> agriculture and forestry mapping and monitoring
> geological and geomorphological mapping
> geodetic mapping and observation of tectonic motion
> hydrological assessment and forecasting
> resource mapping
> hazard and disaster assessment

Oceans
> coastal zone management
> wave forecasting
> measurement of sea-floor topography and determination of the oceanic geoid
> location of fishing areas
> monitoring surface pollution

Cryosphere
> snow monitoring and runoff prediction
> sea ice and icebergs
> glaciers and ice sheets

Aim and scope of this book

This book was conceived with several aims in mind. For the comparative new-comer, it attempts to provide brief definitions of terms and concepts commonly encountered in remote sensing. For the more experienced practitioner, the book collects reference data that are often scattered rather widely throughout the existing literature, and it also aims to provide an insight into the wide range of applications of remote sensing so that specialists in one area can obtain help-ful insights from others.

The book is organised alphabetically, as a dictionary, in order to facilitate the location of information from unfamiliar areas. Extensive cross-referencing should ensure that the required entry can be found even if the user is unsure of the best starting point. The entries cover the following main areas: definitions and surveys of the principal types of remote sensing system; definitions of the concepts defining the operation and performance of remote sensing systems; propagation and scattering of electromagnetic radiation; characterisation of the Earth's atmosphere; the orbital dynamics of satellites; brief notes on the principal space agencies; remote-sensing satellites; spaceborne remote sensing instruments; concepts and techniques of image processing of remotely sensed data. Brief notes on a number of important applications of remote sensing are also included.

The book addresses itself to remote sensing of the Earth's surface and atmo-sphere from space. Airborne systems are excluded, although reference is made to them where appropriate. Similarly, the majority of short-duration Space Shuttle missions are not described. Observation programmes directed primarily towards astronomy or the exploration of the solar–terrestrial environment are also excluded, as are specifically military applications.

A comment is appropriate with regard to the information provided on space agencies. The last few years have seen an increasing trend towards spaceborne missions operated jointly by national (or international) space agencies and by commercial companies, and it seems likely that this trend will continue. The book therefore includes some information on relevant commercial operators; naturally, such inclusion constitutes neither an advertisement for nor an endorsement of the company in question.

While the editor and contributors have made every effort to ensure that the book is as up-to-date as it can be, consistent with publishing timetables, some 'dating' of the material is inevitable as a result of the rapid evolution of national and international Remote Sensing programmes. The book attempts to address this fact by including information on the major satellite programmes and missions planned for the next ten years or so. Some of this information is neces-sarily provisional, and in order to help the user of this book to keep track of changes in mission profiles, relevant World Wide Web (Internet) addresses (URLs, or uniform resource locators) have been included wherever they have been thought helpful. There are also a number of useful WWW pages providing general information on remote sensing, particularly NASA's 'Space Hotlist', CEOS's 'Yellow Pages' and the Centre for Earth Observation's pages. At the time of writing, the URLs of these sites are

http://www.hq.nasa.gov/osf/hotlist/
http://www.smithsys.co.uk/yp/intro.htm
http://ewse.ceo.org/

These and other URLs listed in the book may change over time, as may their content and relevance. The wide range of WWW search engines should enable the user to locate relevant sites in future.

This book necessarily represents a synthesis of data and information from many sources. As such, it does not readily lend itself to the provision of specific references to books and to articles published in scientific journals. A short bibliography has been provided to enable the interested reader to pursue any of the topics discussed in the book in greater depth.

A note about abbreviations and acronyms

Remote sensing is a discipline that tends to generate large numbers of abbreviations and acronyms, especially in the naming of the various instruments carried by remote-sensing satellites. In general, definitions will be found under the appropriate abbreviation (for example, *ATSR*), although the definition will also be cross-referenced from the full name (*along-track scanning radiometer*). The editor believes that this policy places the definition under the more commonly used term; it also avoids the difficulty caused by the fact that, while the abbreviated name of an instrument is generally fixed, the 'expansion' of the abbreviation is not unknown to vary.

Errors, omissions and suggestions

The editor accepts complete responsibility for the content and structure of the book. He would welcome suggestions for improvements.

Gareth Rees
Cambridge

Acknowledgements

The editor gratefully acknowledges the work of the contributors to this book, without which it could not have been written. He also thanks the following people who provided a critical review of the manuscript: Bernard Devereux, Department of Geography, University of Cambridge; Christopher S. M. Doake, British Antarctic Survey; David R. Wilson, University of Cambridge Committee for Aerial Photography; Robin G. Williams, Department of Atmospheric Science, University of Alabama in Huntsville. Thanks for patience and encouragement are due to Catherine Flack of Cambridge University Press and to Christine Rees.

AATSR (Advanced Along-Track Scanning Radiometer) U.K. optical/near infra-red/thermal infrared mechanically scanned imaging radiometer, planned for inclusion on *Envisat*. Wavebands: 0.65, 0.85, 1.27, 1.6, 3.7, 11.0, 12.0 μm. Spatial resolution: 0.5 km (wavelengths up to 1.6 μm), 1.0 km (wavelengths 1.6 μm and above). Swath width: 500 km.

Like the *ATSR*, the AATSR will use a conical scanning technique which provides data from both nadir and 52° forward of nadir. This allows correction of the thermal infrared data for atmospheric emission and absorption effects.

URL: http://envisat.estec.esa.nl/instruments/aatsr/index.html

Ablation See *glaciers*.

Absorption coefficient Term describing the rate at which energy is lost from electromagnetic radiation as it propagates through an absorbing medium. If the intensity (*radiance* etc.) of the radiation propagating in the *x*-direction is *I*, the absorption coefficient γ_a is defined by

$$\gamma_a = -\frac{dI}{I\,dx}$$

and has dimensions of (length)$^{-1}$. For constant γ_a, the solution of this equation is

$$I = I_0 \exp(-\gamma_a x),$$

or equivalently

$$\ln(I) = \ln(I_0) - \gamma_a x,$$

where I_0 is a constant. (See *Lambert–Bouguer law*.) Absorption coefficients are sometimes also expressed in *decibel* per unit length.

See also *radiative transfer equation, refractive index*.

AC Russian broad-band UV/optical/infrared radiometer for Earth radiation budget measurements, carried on *Meteor-1* satellites. Waveband: 0.3–30 μm. Spatial resolution: 50 km. Swath width: 2500 km.

Accumulation See *glaciers*.

Accuracy, classification See *error matrix*.

1

Across-track direction See *range direction*.

Active Microwave Instrument See *AMI*.

Active system A remote sensing system that emits radiation and analyses the returned component, such as a *lidar*, an imaging *radar*, a *radar altimeter* or a *scatterometer*. Compare *passive system*.

ADEOS (Advanced Earth Observing Satellite) Japanese satellite, operated by *NASDA*, launched in August 1996 with a nominal lifetime of 3 years. The satellite's solar power system failed irrecoverably in June 1997. Objectives: Global land, ocean, and atmospheric observations. Orbit: Circular *Sun-synchronous LEO* at 797 km altitude. Period 101 minutes; inclination 98.6°; equator crossing time 10:30 (descending node). *Exactly-repeating orbit* (585 orbits in 41 days). Principal instruments: *AVNIR, ILAS, IMG, NSCAT, OCTS, POLDER, TOMS*. The satellite also has a laser retroreflector array (RIS) for laser ranging from ground stations and for atmospheric absorption measurements.
ADEOS was also known as Midori.

URL: http://www.eorc.nasda.go.jp/ADEOS/

ADEOS II (Advanced Earth Observing Satellite) Japanese satellite, operated by *NASDA*, scheduled for launch in February 1999 with a nominal mission of 5 years. Objectives: Part of IEOS (International Earth Observation System – see *EOS*) for global change studies through WCRP, GEWEX, CLIVER, IGBP, GCOS, and integrated into the *EOS* programme. Orbit: Circular *Sun-synchronous LEO* at 797 km altitude. Period 101 minutes; inclination 98.6°; equator crossing time 10:30 (descending node). Principal instruments: *AMSR, GLI, ILAS-II, POLDER-2, SeaWinds*. The satellite will also carry a data collection package. ADEOS-II was formerly known as **JPOP** (Japanese Polar Platform).

URL: http://titan.eorc.nasda.go.jp/test/GLI/adeos2.html

Adjacency effect Contribution to the *radiance* (optical or near infrared) of a *pixel* by atmospherically scattered radiation originating in nearby pixels. The adjacency effect is mainly due to scattering by atmospheric *haze*, and has a horizontal scale of the order of 1 km.

Advanced Along-Track Scanning Radiometer See *AATSR*.

Advanced Earth Observing Satellite See *ADEOS, ADEOS II*.

Advanced Land Observing Satellite See *ALOS*.

Advanced Microwave Scanning Radiometer See *AMSR*.

Advanced Microwave Sounding Unit See *AMSU/MHS, AMSU-B*.

Advanced Millimetre-wave Atmospheric Sounder See *AMAS*.

Advanced optical and Near Infrared Radiometer See *AVNIR*.

Advanced SAR See *ASAR*.

Advanced Spaceborne Thermal Emission and Reflection Radiometer See *ASTER*.

Advanced Synthetic Aperture Radar See *ASAR*.

Advanced TIROS-N See *NOAA-6* to *NOAA-8*.

Advanced Very High Resolution Radiometer See *AVHRR*.

Advanced Vidicon Camera Subsystem See *AVCS*.

Advanced Wind Scatterometer See *ASCAT*.

AEM-1 See *HCMM*.

AEM-2 (Applications Explorer Mission 2) U.S. satellite, operated by *NASA*, launched in February 1979 and terminated in November 1981. Objectives: Monitoring of stratospheric gases and aerosols. Orbit: Nominally circular *LEO* at 550 km altitude. Inclination 55°; period 97 minutes. Principal instruments: *SAGE-I*.

Aerosol A suspension of very small (typically 1 nm to 10 μm) solid particles (e.g. dust, sulphates) or liquid droplets in air, mostly found in the atmospheric *boundary layer*. Aerosols are climatically important as a result of the attenuation of solar radiation and their role in *cloud* formation. Some aerosols are also chemically active (see *chemistry, atmospheric*). Scattering and absorption of optical and near infrared radiation by aerosols is the major source of uncertainty in *atmospheric correction*. Concentrations of aerosols are difficult to measure, especially in the *troposphere* and over land, where the high surface reflectance masks the radiation scattered by the aerosol.

Aerosols of different origin have markedly different size distributions and refractive indices, and hence different absorption and scattering properties. The figure summarises the size distributions of rural, urban, tropospheric, maritime, meteoric, stratospheric and volcanic aerosols. The horizontal axis shows the particle size in μm; the vertical axis the value of dN/dr where N is the concentration of particles of size r. All the graphs have been normalised to a total concentration of 1.

Over a limited range of wavelength, both the scattering and absorption coefficients of an aerosol can be described approximately by the *Ångström*

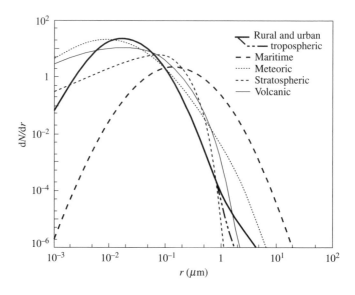

relation

$$\gamma = \gamma_0 (\lambda/\lambda_0)^{-\nu},$$

where λ is the wavelength and γ has the value γ_0 at some reference wavelength λ_0. ν is the Ångström exponent. The table below summarises the approximate values of γ_0 (in km^{-1}) and ν for the optical/near infrared region, corresponding to a total attenuation coefficient of $1\,km^{-1}$ and taking $\lambda_0 = 0.55\,\mu m$.

Aerosol	Total γ_0	ν	Scattering γ_0	ν	Absorption γ_0	ν
rural	1.0	1.3	0.95	1.3	0.05	0.5
urban	1.0	1.2	0.77	1.3	0.23	0.8
tropospheric	1.0	1.8	0.97	1.8	0.03	0.8
maritime	1.0	0.2	0.99	0.2	0.01	−0.1
meteoric	1.0	0.4	0.995	0.4	0.005	−2
stratospheric	1.0	2.1	1.0	2.1	10^{-7}	−6
volcanic	1.0	1.2	0.94	1.2	0.06	1.2

The attenuation coefficient at sea level typically ranges between 0.05 and $0.5\,km^{-1}$; at the tropopause it typically ranges between 0.001 and $0.003\,km^{-1}$.

Measurement of the vertical profile of aerosol scattering coefficient can be made by optical/near infrared limb sounding (e.g. *GOMOS*, *HiRDLS*, *ILAS*, *POAM-2*, *SAGE III*, *SAM-II*), backscatter lidar (e.g. *Alissa*, *GLAS*) or nadir ultraviolet/optical/infrared spectrometry (e.g. *MISR*, *MODIS-N*, *MOS*, *POLDER*).

Agenzia Spaziale Italiana The Italian space agency. See *CEOS*.

4

AIRS (Atmospheric Infrared Sounder) U.S. optical/infrared grating spectrometer, planned for inclusion on *EOS-PM* satellites. Wavebands: 2300 bands between 3.74 and 15.4 µm (spectral resolution 3 nm to 13 nm), 6 bands between 0.4 and 1.7 µm. Spatial resolution: 13.5 km horizontal (at nadir), 1 km vertical. Swath width: 1650 km (scans to ±49° from nadir).

AIRS will provide atmospheric temperature profiles by measuring thermal emission.

URL: http://www-airs.jpl.nasa.gov/

Albedo The fraction of incident radiation that is reflected by a surface. See *bidirectional reflectance distribution function*.

Albedo, planetary The fraction of incident solar radiation that is reflected back into space by a planet. The Earth's planetary albedo (approximately 30%) is a fundamental factor in determining its energy budget and hence climate (see *Earth radiation budget*). It can be measured by calibrated optical/near infrared radiometry from *LEO* or from *geostationary* orbit. See *bidirectional reflectance distribution function*.

Algorithm A mathematical process connecting a remote sensing measurement to the surface or atmospheric property of interest.

Aliasing A periodic phenomenon of frequency f_1, sampled at a frequency f_0, will appear to vary with a frequency f_a given by

$$ f_a = f_1 - f_0 \left[\frac{f_1}{f_0} + \frac{1}{2} \right], $$

where $[x]$ is the least-integer function. This phenomenon is called aliasing, and f_a is the aliased frequency. See also *Nyquist frequency*.

Alissa (l'Atmosphere par Lidar sur Saliout) French backscatter *lidar*, for measurement of cloud and aerosol structure, carried on *Mir-1*. Wavelength: 532 nm (Nd-YAG laser). Spatial resolution: 300 m (horizontal), 150 m (vertical). Pulse repetition frequency: 8 Hz.

Almaz-1 Russian satellite, launched in March 1991, lifetime 1.5 years. Objectives: Global land, ocean, atmosphere observations. Orbit: Nominally circular *LEO* at 270–380 km altitude. Period 92 minutes; inclination 73°. Principal instruments: *SAR*, *UHF radiometer*. 'Almaz' is the Russian word for 'diamond'.

Almaz-1 B Russian satellite, scheduled for launch in 1998 with a nominal lifetime of 3 years. Objectives: Global land, ocean, atmosphere observations. Orbit: Circular *LEO* at 400 km altitude. Period 90 minutes; inclination 73°. Principal instruments: *Balkan-2*, *MSU-E*, *MSU-SK*, *OEA*, *SAR-3/SLR-3*, *SAR-10*, *SAR-70*, *SROM*.

Along-track direction The direction on the Earth's surface parallel to the motion of a *side-looking radar* or *synthetic aperture radar*, also called the azimuth direction.

Along-Track Scanning Radiometer See *ATSR*.

ALOS (Advanced Land Observing Satellite) Japanese satellite planned for launch in 2002. Objectives: Cartography, environmental monitoring, hazard monitoring. Orbit: Circular *Sun-synchronous LEO*. Equator crossing time 10:30 descending. Principal instruments: *AVNIR-2, VSAR*.

The ALOS mission was formerly called HIROS.

URL: http://www.goin.nasda.go.jp/GOIN/NASDA/act/alos.html

ALT U.S. dual-frequency nadir-viewing radar altimeter, carried on *Topex-Poseidon*. Frequencies: K_u band (13.6 GHz) and C band (5.3 GHz). Pulse length (uncompressed): 102 µs; (compressed): 3.1 ns. Range precision: 2.4 cm. Beam-limited footprint: 26 km (13.6 GHz); 65 km (5.3 GHz). Pulse-limited footprint: 2.2 km.

Dual frequency operation allows for correction of ionospheric delays.

URL: http://www-aviso.cls.cnes.fr/English/TOPEX_POSEIDON/More_On_Payload.html

Altimetry Microwave Radiometer See *AMR*.

Altitude, orbital See *height, orbital*.

AMAS (Advanced Millimetre-wave Atmospheric Sounder) European passive microwave atmospheric limb sounder, operating in the frequency range 298–626 GHz (wavelengths 0.5–1.0 mm), planned for inclusion on *Meteor-3M* satellite. Frequencies: 298, 301, 302, 325, 346, 500, 501, 626 GHz. Spatial resolution: vertical: 1 km (troposphere); 3 km (stratosphere); 5–10 km (mesosphere). Horizontal: 300 km. Height range: 5–100 km.

AMAS is designed to measure temperature and pressure profiles, ozone, water vapour, N_2O and other chemical constituents of the middle and upper atmosphere.

AMI (Active Microwave Instrument) European *synthetic aperture radar*/microwave *scatterometer* carried on *ERS-1* and *-2* satellites. Frequency: C band (5.3 GHz). Polarisation: VV. Incidence angle: 23° at mid-swath. Spatial resolution: 30 m (wave and image modes), 50 km with data on a 25 km grid (scatterometer mode). Swath width: 5 km × 5 km (wave mode), 100 km (image mode), 500 km (scatterometer mode). Radiometric resolution: 2.5 dB in image mode, 0.3 dB in scatterometer mode.

In scatterometer mode, the AMI generates beams in directions 45°, 90°, 135° (zero is along-track, 90° to the right). The near edge of the swath is 150 km to the right of the sub-satellite track; the far edge is 650 km from it. Wind velocity vectors over oceans can be retrieved with an accuracy of $\pm 2\,\mathrm{m/s}$, $\pm 20°$, in the range 4 to $24\,\mathrm{m/s}$.

The duty cycle for image mode is approximately 12% (i.e. 12 minutes of data can be stored per orbit). For wave and scatterometer modes the duty cycle is 100%.

URL: http://earth1.esrin.esa.it/f/eeo2.402/ERS1.3

Among-class covariance See *canonical components*.

AMR (Altimetry Microwave Radiometer) U.S. nadir-viewing *passive microwave radiometer*, planned for inclusion on *TPFO*. Frequencies: 18.2, 23.8, 34.0 GHz. Spatial resolution: 29 km, 23 km, 16 km respectively.

The AMR will be used to provide propagation delay estimates for correction of data from the *DFA*. It is based on the design of the *TMR*.

AMSR (Advanced Microwave Scanning Radiometer) Japanese conically scanned *passive microwave radiometer*, planned for inclusion on *ADEOS II*. Frequencies: 6.9, 10.65, 18.7, 23.8, 36.5, 50.3, 52.8, 89 GHz. Polarisation: H and V. Spatial resolution: 50 km at 6.9 GHz to 5 km at 89 GHz. Swath width: 1600 km. Sensitivity: 0.3–1 K. Absolute accuracy: 1–2 K.

AMSU/MHS (Advanced Microwave Sounding Unit/Microwave Humidity Sounder). U.S.–European mechanically scanned *passive microwave radiometer* for profiling atmospheric temperature and water vapour, planned for inclusion on *EOS-PM*, *Metop* and future *NOAA* satellites. Channels 1 and 2 are the

Channel	Frequency (GHz)	Bandwidth (MHz)
1	23.80	280
2	31.40	180
3	50.30	180
4	52.80	400
5	53.60 ± 0.12	170
6	54.40	400
7	54.94	400
8	55.50	330
9	57.29	330
10	57.29 ± 0.22	78
11	57.29 ± 0.32	36
12	57.29 ± 0.32	16
13	57.29 ± 0.32	8
14	57.29 ± 0.32	3
15	89.00	6000
16	89.00	2800
17	157.0	2800
18	183.31 ± 1	500
19	183.31 ± 3	1000
20	183.31 ± 7	2200

AMSU-A2 instrument, channels 3–15 the AMSU-A1, and channels 16–20 the MHS unit.

Spatial resolution: 40 km at nadir (AMSU); 15 km at nadir (MHS). Swath width: 1650 km. Height range: 0–70 km. Temperature resolution: 0.25–1.2 K (AMSU-A), 1.0 K (MHS).

AMSU uses channels 3–14 for temperature profiling, using the oxygen absorption band between 50 and 60 GHz. The other AMSU channels provide corrections for surface temperature and atmospheric water and water vapour. MHS, which is similar to *AMSU-B*, provides water vapour profiles, and will also give qualitative estimates of precipitation rates.

AMSU-B (Advanced Microwave Sounding Unit-B) U.K. mechanically scanned passive microwave radiometer for atmospheric water vapour profiling, planned for inclusion on future *NOAA* satellites.

Channel	Frequency (GHz)	Bandwidth (GHz)
1	89.00	6
2	150.0	4
3	183.31 ± 1	1
4	183.31 ± 3	2
5	183.31 ± 7	4

Spatial resolution: 16 km at nadir. Swath width: 2200 km. Height range: 0–42 km. Temperature resolution: 1.0–1.2 K.

AMSU-B is very similar to MHS (see *AMSU-A/MHS*).

Ångström relation The assumption that the *optical thickness* τ of an atmospheric *aerosol* varies with wavelength λ according to

$$\tau \propto \lambda^{-\nu},$$

where ν is the Ångström exponent. In the absence of any other information, it is often assumed that $\nu = 0$.

Angular frequency see *frequency*.

Antenna A transducer that couples a guided electromagnetic wave on a transmission line to an unguided wave in the medium surrounding the antenna, and vice versa. Antennas are used by both passive microwave radiometers (see *passive microwave radiometry*) and *radar* systems. The directional pattern of an antenna, specified by its *power pattern*, consists of a main beam surrounded by a number of sidelobes, with the effective width of the beam β (radians) in a given plane being approximately equal to λ/D, where λ is the wavelength and D is the width of the antenna in that plane (see *beamwidth*). The antenna may consist of a single element, such as a parabolic dish, a horn, or a microstrip radiating patch, or it may consist of an array of such elements. A synthetic aperture may be formed by combining the signals received by a real antenna as it travels in space and then processing them together as if they had been received by a long array of individual elements, thereby attaining high angular resolution in the plane containing the synthetic array. This is the basis of *synthetic aperture radar*.

Antenna temperature The antenna temperature represents the power received by an *antenna* and transferred to the receiver. For a lossless antenna with a *power pattern* consisting of only a single main beam with no sidelobes, the antenna temperature T_A is equal to the *brightness temperature* T_b incident upon the antenna through its main beam, where T_b represents the power radiated by the scene observed by the antenna. A real antenna, however, is characterised by a radiation efficiency η_l and a main-beam efficiency η_m, where η_l accounts for ohmic losses in the antenna structure and η_m accounts for the fraction of the total antenna power pattern contained in its main beam. In that case, T_A is related to T_b through

$$T_A = \eta_l \eta_m T_b + \eta_l (1 - \eta_m) T_{SL} + (1 - \eta_l) T_0,$$

where T_{SL} is the average brightness temperature of radiation incident upon the antenna along directions other than through its main beam and T_0 is the physical temperature of the antenna. It is possible to achieve values of η_l as high as 0.99. It is also possible to design the antenna such that its side-lobe levels are very low, thus achieving values of η_m in the range 0.90 to 0.98, but this is accomplished at the expense of widening the main beam by as much as a factor of two, compared with the narrowest beam attainable for an antenna of a given size, with a corresponding loss in angular resolution.

Apogee The furthest point of a satellite's *orbit* from the Earth's centre.

Applications Explorer Mission See *AEM-2, HCMM*.

APT (Automatic Picture Transmission)
1. See *AVHRR*.
2. U.S. visible-wavelength *vidicon* imaging system with direct transmission of image data, carried on even-numbered *ESSA* satellites, *Nimbus-1, TIROS-8* and *NOAA-1*. Waveband: 0.45–0.65 μm. Spatial resolution: 4 km at nadir. Swath width: 1200–1600 km (3100 km from ESSA).

Archaeological site detection The mastery of nature by man results in a transformation of the landscape through agriculture and construction. In arid parts of the world it is often easy to see the remains left by human activity which survive for long periods above the ground. But in heavily occupied and farmed regions most surface traces have been obliterated. Traces remaining in the ground just below the surface are usually studied through excavation, a tool of last resort, since it destroys the evidence. Techniques for non-destructive investigation have been developed to protect sites by leaving them undisturbed.

The methods derive from geological geophysical prospecting and low altitude aerial survey. Buried remains are detected and mapped by measuring or recording differences in water content, magnetic, thermal and mechanical properties, and displaying these with modern processing methods as two- or three-dimensional coloured images. Low level oblique aerial photographs have been used since the 1920s and are still the most economical approach. Electrical, magnetic, electromagnetic, *radar* and sensitive remote temperature measurements add techniques for use under widely varying geological, climatic and man-made conditions. Buried structures are recognised by skilled archaeologists via their characteristic shapes in properly processed images made from such data.

Argon
1. A minor constituent of the Earth's *atmosphere*.
2. Series of U.S. military reconnaissance satellites, operating between February 1961 and August 1964. The satellites carried panchromatic cameras (Keyhole) with a resolution of 140 m. The data are now declassified.

URL: http://edcwww.cr.usgs.gov/dclass/dclass.html

Argos French/U.S. data collection system, carried on *TIROS-N, NOAA-6* onwards. Argos-equipped spacecraft carry a Data Collection and Location System (DCLS or DCS) package capable of receiving transmissions (402 MHz) from registered data-collection platforms (on ships, buoys, aircraft etc.). These transmissions are downlinked to ground stations for processing (position determination and data formatting). Positions can be determined to ±300 m (±30 m on some days).

ASAR (Advanced SAR) European *synthetic aperture radar* planned for inclusion on *Envisat*. Frequency: C band (5.3 GHz). Polarisation: VV and HH. Incidence angle: 15° to 45° at mid swath (image, wave and alternating polarisation modes). Spatial resolution: 30 m (image, wave and alternating polarisation modes), 100 m (wide swath mode), 1000 m (global monitoring mode). Swath width: 5 km × 5 km images (wave mode), 56–120 km with 7 sub-swaths (image and alternating polarisation modes), 400 km with 5 sub-swaths (wide swath and global monitoring modes). Radiometric resolution: 1.3 to 1.7 dB.
    All modes can operate at either HH or VV polarisation. The 'alternating polarisation' mode interleaves observations in each polarisation state.

URL: http://envisat.estec.esa.nl/instruments/asar/index.html

ASCAT (Advanced Wind Scatterometer) European microwave wind *scatterometer* planned for inclusion on *Metop*. Frequency: C band (5.3 GHz). Polarisation: VV. Beam angles: [0° = forward, 90° = right] ±45°, ±90°, ±135°. Spatial resolution: 50 km (data provided on 25 km grid). Swath width: two 500 km swaths (left and right sides). Near edges 150 km from sub-satellite track; far edges 650 km. Accuracy: ±2 m/s, ±20° for ocean surface wind vectors.
    ASCAT is derived from the *AMI* scatterometer.

Ascending node The point in a satellite's *orbit* at which it crosses the Earth's equatorial plane from the southern to the northern hemisphere.

ASI (Agenzia Spaziale Italiana) See *CEOS*.

ASTER (Advanced Spaceborne Thermal Emission and Reflection Radiometer) Japanese optical/near infrared/thermal infrared multispectral imaging radiometer, to be carried on *EOS AM-1* satellite. Wavebands and spatial resolution:

Band	Waveband (μm)	Spatial resolution (m)
1	0.52–0.60	15
2	0.63–0.69	15
3N*	0.76–0.86	15
3B*	0.76–0.86	15
4	1.60–1.70	30
5	2.15–2.19	30
6	2.19–2.23	30
7	2.24–2.29	30
8	2.30–2.37	30
9	2.36–2.43	30
10	8.13–8.48	90
11	8.48–8.83	90
12	8.93–9.28	90
13	10.25–10.95	90
14	10.95–11.65	90

*Band 3 views both towards nadir and backwards, to give stereoscopic viewing.

Swath width: 60 km, steerable to ±100 km across-track from sub-satellite track (±300 km for optical/near infrared bands).

ASTER was formerly known as **ITIR** (Intermediate Thermal Infrared Radiometer).

URL: http://asterweb.jpl.nasa.gov/asterhome/

ATMOS U.S. limb-sounding infrared spectrometer, for molecular profiling of the atmosphere, carried on *Spacelab-3*. Waveband: 2.2 to 16 μm (Fourier transform spectrometer).

Atmosphere
1. A unit of pressure, equal to 101.325 kPa.
2. The gaseous envelope surrounding the Earth. The main constituents of the atmosphere are nitrogen, oxygen, water vapour, argon and carbon dioxide, though other gases present at low concentrations can also have a significant effect on the *Earth radiation budget*. The composition and physical properties of the atmosphere vary considerably with height, and also with geographical position (especially with latitude, but also with proximity to land or sea, or

to industrial or rural areas). The table below shows the mean composition of the atmosphere at sea level, in terms of the fraction by volume, and the mean column-integrated quantity of gases, expressed in moles per square metre.

Gas	Volume fraction	Column integral (moles per m^2)
N_2	0.7808	2.784×10^5
O_2	0.2095	7.468×10^4
H_2O	$10^{-3} - 2.8 \times 10^{-2}$	$3.6 \times 10^2 - 9.8 \times 10^3$
Ar	9.34×10^{-3}	3.32×10^3
CO_2	3.5×10^{-4}	1.3×10^2
Ne	1.8×10^{-5}	6.5
He	5.2×10^{-6}	1.9
CH_4	1.8×10^{-6}	0.62
Kr	1.1×10^{-6}	0.41
CO	$6 \times 10^{-8} - 10^{-6}$	0.02–0.4
SO_2	10^{-6}	0.4
H_2	5×10^{-7}	0.2
O_3	$10^{-8} - 10^{-6}$	0.11
N_2O	2.7×10^{-7}	0.09
Xe	9×10^{-8}	0.03
NO_2	$5 \times 10^{-10} - 2 \times 10^{-8}$	$2 \times 10^{-4} - 9 \times 10^{-3}$

The atmosphere is conventionally divided into a number of layers on the basis of its vertical temperature structure. The **troposphere**, in which the temperature usually decreases with height, extends from sea level to a height of about 10 km. This is the most turbulent layer, in which most meteorological phenomena occur. The lowest region of the troposphere, the *boundary layer*, is approximately 1 km deep. Above the troposphere is the **stratosphere**, in which the temperature increases with height (as a result of increasing absorption of solar ultraviolet radiation by ozone), extending up to about 50 km. The molecular concentration of ozone increases with height in the stratosphere, reaching a maximum at about 50 km. Between about 50 km and 85 km is the **mesosphere**, in which the temperature decreases with height, and above this is the **thermosphere**, in which it increases. Figure A summarises the variations of temperature, pressure and density between sea level and the lower part of the thermosphere, although it is representative only since the heights of the layer boundaries vary considerably with latitude.

Absorption of electromagnetic radiation by the atmosphere limits the range of wavelengths or frequencies at which remote sensing of the Earth's surface can be performed. Conversely, observations made at the wavelength of an absorption line give the possibility of *atmospheric sounding* of temperature, pressure and molecular composition of the atmosphere. Figures B and C summarise the typical atmospheric attenuation spectrum for a one-way vertical path through the atmosphere, although these values are again dependent on latitude and on the concentrations of different molecular species.

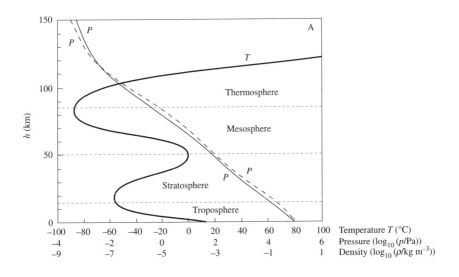

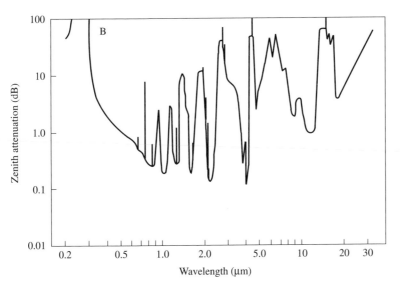

Figure B shows schematically the zenith attenuation plotted against wavelength in the ultraviolet, optical and infrared parts of the electromagnetic spectrum. The main peaks are as follows (µm):

CH_4:	1.66	2.1	2.3	2.4	3.3			
CO_2:	1.95	2.0	2.1	2.7	2.8	3.3	4.3	15
H_2O:	0.72	0.82	0.93	1.12	1.37	1.85	2.6	5.9	6.5
N_2O:	3.8	7.8						
O_3:	0.26	0.60	4.7	9.7				
O_2:	0.69	0.76	1.25					

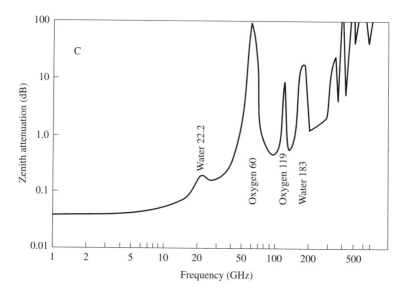

Figure C shows schematically the zenith absorption plotted against frequency in the microwave spectrum. The absorption below 20 GHz is due to non-resonant absorption by oxygen. The peaks above 300 GHz are due to water vapour.

l'Atmosphere par Lidar sur Saliout See *Alissa*.

Atmospheric chemistry See *chemistry, atmospheric*.

Atmospheric correction The useful information about a target area of the land, sea or clouds is contained in the physical properties of the radiation that leaves that target area, whereas what is measured by a remote sensing instrument are the properties of the radiation that arrives at the instrument. It is therefore necessary to correct the satellite- (or aircraft-) received data to allow for the effects on the radiation as it passes through the atmosphere. In theory the passage of the radiation through the atmosphere is described by the *radiative transfer equation*; however, the values of the various parameters that appear in the radiative transfer equation are not known sufficiently accurately to make direct and explicit solution of the radiative transfer equation a feasible approach. Thus in practice more empirical methods are used depending, among other things, on the wavelength of the radiation concerned.

For **microwave** radiation, at the higher microwave frequencies the atmosphere is an absorbing medium which attenuates the energy emitted from the surface. The atmosphere also emits energy as described by *Kirchhoff's law*. If we consider the atmosphere as a stratified medium in which the temperature and the microwave absorption coefficient are functions of height, then it is possible to represent the temperature describing the radiation received at a sensor mounted on a spacecraft as consisting of the following four components: (i) a contribution arising from sky radiation which has been reflected at the surface of the Earth and attenuated by its passage upwards through the atmosphere; (ii) the down-welling emitted radiation from the atmosphere which has been reflected at the surface of the Earth and attenuated by its passage upwards through the atmosphere; (iii) the radiation emitted from the surface in the direction towards the spacecraft and attenuated by its passage upwards through the atmosphere and (iv) the upwelling radiation emitted by the atmosphere. It is the third contribution which constitutes the required signal; therefore it is necessary to estimate the contributions from (i), (ii) and (iv) in order to determine the temperature of the surface of the Earth from the signal generated at the instrument on board the spacecraft.

For **thermal infrared** radiation, i.e. relatively long wavelength infrared radiation that is emitted from the surface of the Earth, the major effects of the atmosphere arise from the water vapour. There are three main methods for the atmospheric correction of thermal infrared data.

(i) The use of the *LOWTRAN* computer package that requires as input an atmospheric profile of pressure, temperature and humidity. While one can use a model, or average locational and seasonal, atmosphere as input to the LOWTRAN computer package, this is not particularly satisfactory because the water vapour content of the atmosphere is highly variable, both spatially and temporally, and it is really necessary to use the values of the parameters that apply to the actual atmospheric conditions at the time that the remotely sensed data were gathered. Such a profile can be obtained either from radiosonde data or from a satellite-flown sounding instrument (e.g. the *TOVS* instrument).

(ii) The use of the split-window method to eliminate the atmospheric effects between two channels of nearby, but different, wavelength ranges. In this method the brightness temperature T_{bi} ($i = 1, 2$), that is the equivalent black-body temperature of the radiation reaching the spacecraft in each of two spectral channels (bands) ($i = 1, 2$) is first calculated using the calibration data for the infrared channels of the scanner. The equivalent black-body temperature, T_0, for the radiation leaving the surface of the Earth, is then determined as a linear combination of these two brightness temperatures:

$$T_0 = A_0 + A_1 T_{b1} + A_2 T_{b2},$$

where the coefficients A_0, A_1 and A_2 have been determined empirically and are also assumed to be of general validity.

(iii) The use of a two-look method, e.g. by a conical scanning system such as that in the Along Track Scanning Radiometer (*ATSR*) flown on ERS-1 and ERS-2. Again two brightness temperatures are determined for the two views of a given area on the surface of the Earth and a similar relation to that in (ii) is used to determine T_0, but with its own set of values of A_0, A_1 and A_2.

For **optical and near infrared** wavelength radiation the atmospheric corrections are particularly important, in the sense that they are much larger, as a
percentage of the surface-leaving radiation, than is the case for microwave or
thermal infrared radiation. When an electromagnetic wave strikes a particle,
no matter the size of the particle, a part of the incident energy is scattered in
all directions. This scattered energy is called diffuse radiation. An expression
for the energy scattered by spherical particles can be obtained theoretically
by solution of Maxwell's equations of electromagnetism. By using the parameter $\alpha = \pi D/\lambda$, where D is the mean diameter of the scattering particles
and λ is the wavelength of the incident radiation, we can distinguish the following cases:

(i) $\alpha < 0.1$. In this case we have *Rayleigh scattering*. This is applicable to the
scattering of solar radiation by air molecules, of which the majority have
size of the order of 0.1 nm. The Rayleigh scattering is not very sensitive
to changes in atmospheric conditions.
(ii) $0.1 < \alpha < 50$. In this case we have *Mie scattering*. This kind of scattering is
applied to the scattering of solar radiation by particles with size greater
than 10 nm like *aerosols*.

The simplest atmospheric correction algorithms for optical and near infrared
observations assume that the darkest *pixels* in an image (usually arising from
deep shadows or from areas of deep clear water) correspond to zero *radiance*,
and that the effect of atmospheric scattering of radiation into the line of sight is
purely additive, so that the radiance detected from the darkest pixels can simply
be subtracted from all pixels in the image.

More sophisticated algorithms use models of atmospheric absorption and
scattering. These algorithms take into account five contributions to the radiance
measured at the satellite: (i) radiation scattered directly from the surface into the
sensor; (ii) radiation scattered directly from the atmosphere into the sensor; (iii)
radiation scattered from the surface and then diffusely by the atmosphere into
the sensor; (iv) radiation scattered diffusely by the atmosphere and then directly
from the surface into the sensor; (v) radiation multiply scattered from the
surface and atmosphere. The at-satellite radiance L can be expressed as an
equivalent reflectance ρ^*, defined as

$$\rho^* = \frac{\pi L}{\mu_0 E},$$

where E is the exoatmospheric *irradiance* and μ_0 is the cosine of the solar zenith
angle. This equivalent reflectance can then be written as the sum of terms representing the contributions (i) to (v) above. In turn, these are:

(i) $\exp(-\tau[\mu_0^{-1} + \mu^{-1}])\rho,$

where μ is the cosine of the incidence angle, ρ is the value of the *bidirectional
reflectance distribution function* of the surface appropriate to the viewing geometry, and τ is the *optical thickness* of the atmosphere for a vertical path (see
Lambert–Bouguer law).

(ii) ρ_a,

the atmospheric path reflectance, which is also a function of the viewing geometry.

(iii) $t_d(\mu)[\exp(-\tau/\mu) + t_d(\mu_0)]\langle\rho\rangle$,

where t_d is the diffuse transmittance of the atmosphere and $\langle\rho\rangle$ is the average reflectance of that region of the surface, adjacent to the target pixel, from which radiation is scattered into the sensor.

(iv) $t_d(\mu_0)\exp(-\tau/\mu)\bar{\rho}$,

where $\bar{\rho}$ is the average surface reflectance under diffuse illumination.

(v) $\dfrac{\langle\rho\rangle s[\exp(-\tau/\mu_0) + t_d(\mu_0)][\exp(-\tau/\mu)\bar{\rho} + t_d(\mu)\langle\rho\rangle]}{1 - \langle\rho\rangle s}$,

where s is the spherical albedo of the atmosphere.

The greatest uncertainty in these algorithms is the effect of atmospheric aerosols.

Atmospheric Infrared Sounder See *AIRS*.

Atmospheric sounders The table summarises the characteristics of the main optical and infrared nadir-looking or scanning instruments used for *atmospheric sounding*. Several instruments have been carried on more than one satellite: columns 2 and 3 show the first. Column 4 (waveband) uses the following abbreviations: U = ultraviolet, V = visible, N = near infrared, T = thermal infrared. Column 6 (applications) uses the following abbreviations: T = temperature, W = water vapour, O = ozone, M = other molecular species. See also *electro-optical sensors, imaging radiometer, limb-sounding, multispectral imager, passive microwave radiometry*.

Instrument	Satellite	Launch	Waveband	Type	Applications
IRIS	Nimbus-3	1969	T	Nadir spectrometer	T
SIRS	Nimbus-3	1969	T	Nadir radiometer	T
BUV	Nimbus-4	1970	U	Scanning spectrometer	O
ITPR	Nimbus-5	1972	T	Nadir radiometer	T
HIRS	Nimbus-6	1975	NT	Scanning radiometer	TW
LRIR	Nimbus-6	1975	T	Limb radiometer	O
SSD	DMSP 5D-1	1976	U	Limb radiometer	O
SSH	DMSP 5D-1	1976	T	Scanning radiometer	WOM
SI-GDR	Meteor-Priroda	1976	T	Nadir spectrometer	T
LIMS	Nimbus-7	1978	T	Limb spectrometer	TWOM
SAM-II	Nimbus-7	1978	N	Limb radiometer	M
SAMS	Nimbus-7	1978	T	Nadir radiometer	TWM
TOMS	Nimbus-7	1978	U	Scanning radiometer	OM
SSU	TIROS-N	1978	T	Nadir radiometer	T
SBUV	Nimbus-7	1978	U	Scanning spectrometer	O
HIRS/2	TIROS-N	1978	T	Scanning radiometer	TW
SAGE-I	AEM-2	1979	UVN	Limb radiometer	OM
SAGE-II	ERBS	1984	UVN	Limb radiometer	OM
ATMOS	Spacelab	1985	NT	Limb spectrometer	M
MKS-M-AS	Mir-1	1986	N	Nadir spectrometer	T

Instrument	Satellite	Launch	Waveband	Type	Applications
CLAES	UARS	1991	T	Limb spectrometer	WOM
HALOE	UARS	1991	T	Limb spectrometer	M
ISAMS	UARS	1991	T	Nadir radiometer	WOM
POAM-2	SPOT-3	1993	UVN	Limb spectrometer	TWOM
GOES sounder	GOES-Next	1994	T	Scanning radiometer	TWO
GOME	ERS-2	1995	UVN	Nadir spectrometer	OM
ILAS	ADEOS	1996	NT	Limb spectrometer	TWOM
IMG	ADEOS	1996	T	Nadir spectrometer	M
Ozon-M	Mir-1	1996	UVN	Limb spectrometer	OM
OSIRIS	ODIN	1998	UVN	Limb spectrometer	M
MOPITT	EOS-AM	1998	T	Scanning spectrometer	M
GOMOS	Envisat	1998	UVN	Limb spectrometer	TWOM
MIPAS	Envisat	1998	T	Limb spectrometer	M
Sciamachy	Envisat	1998	UVN	Nadir (+limb) spectrometer	M
SAGE-III	Meteor-3M	1999	UVN	Limb spectrometer	TOM
SFOR	Meteor-3M	1999	U	Nadir spectrometer	O
AIRS	EOS-PM	2000	VNT	Scanning spectrometer	T
IASI	Metop	2000	T	Nadir spectrometer	TWOM
HIRS/3	Metop	2000	VNT	Scanning radiometer	TWOM
OMI	Metop	2000	UVN	Nadir spectrometer	O
HiRDLS	EOS-Chem	2002	T	Limb spectrometer	TWOM
ODUS	EOS-Chem	2002	U	Nadir spectrometer	O
TES	EOS-Chem	2002	NT	Nadir (+limb) spectrometer	WOM

Atmospheric sounding The goal of atmospheric sounding is to obtain profiles of temperature and gas concentration (including atmospheric density through the concentration of molecular oxygen) through the troposphere and stratosphere from upwelling radiation emitted by the atmosphere. Sounding instruments measure the radiation at a number of wavelengths where most of the radiation has been emitted by the atmosphere itself (see *atmospheric sounders*). By looking at wavelengths in which the atmosphere has stronger or weaker transmission of radiation the data from different wavelengths can be used to look at various levels of the atmosphere (see *radiative transfer equation*). With a knowledge of the distribution of the gases it is possible to invert these measurements from satellite sensors to retrieve the temperature profile or, if the temperature structure is known, the distribution of the gases. Both these problems are ill founded and more information is required before they can be solved, the data coming from climatology or a numerical model of the atmosphere. Sounding takes place with data from infrared or microwave radiometers on both polar *LEO* and *geostationary* satellites.

Currently, sounding is carried out using a combination of infrared and microwave measurements with the infrared data providing soundings in cloud-free or partly cloudy conditions and microwave observations giving sounding where there is thick cloud. Operational meteorological sounding is carried out from the polar orbiting satellites using nadir viewing instruments that scan across a 1000–2000 km swath. A number of sensors on research satellites in polar orbit have used a *limb-sounding* technique where an oblique view of the Earth's atmosphere is obtained. Such a technique allows *temperature, water vapour* and minor constituents to be retrieved.

See also *chemistry, atmospheric* and *ozone.*

ATN See *NOAA-6* to *-8.*

ATSR (Along-Track Scanning Radiometer) U.K. near infrared/thermal infrared mechanically scanned imaging radiometer, carried on *ERS-1.* Wavebands: 1.6, 3.7, 11.0, 12.0 μm. Spatial resolution: 1.0 km. Swath width: 500 km.

The ATSR uses a conical scanning technique which provides data from both nadir and 52° forward of nadir. This allows correction of the thermal infrared data for atmospheric emission and absorption effects.

The ATSR also includes a nadir-viewing passive microwave radiometer (two channels, at 23.8 and 36.5 GHz) with a spatial resolution of 20 km. The principal function of this is to determine the column-integrated atmospheric liquid water and water vapour content for correction of *SST* measurements and radar altimeter ranges.

ERS-2 carries the ATSR-2 instrument, which has three optical/near infrared wavebands, at 0.555, 0.659 and 0.865 μm, in addition to the four bands of the ATSR instrument.

URL: http://earth1.esrin.esa.it/f/eeo2.402/ERS1.5

Attenuation Reduction in the intensity of a parallel beam of electromagnetic radiation as a result of both scattering and absorption. The attenuation coefficient γ is defined such that, if thermal emission of radiation is negligible, the intensity I of the forward-propagating radiation obeys the differential equation

$$\frac{\mathrm{d}I}{\mathrm{d}x} = -\gamma I,$$

where x is the distance measured in the propagation direction. The attenuation coefficient γ is given by

$$\gamma = \gamma_\mathrm{a} + \gamma_\mathrm{s},$$

where γ_a is the absorption coefficient and γ_s is the scattering coefficient. The **attenuation length**, defined as $1/\gamma$, is the distance over which the intensity of the radiation is reduced by a factor of e.

Attenuation is also called **extinction**. See *absorption coefficient, Lambert–Bouguer law, radiative transfer equation, scattering.*

Automatic linear enhancement See *contrast enhancement.*

Automatic Picture Transmission See *APT.*

AVCS (Advanced Vidicon Camera Subsystem) U.S. visible-wavelength *vidicon* imaging radiometer carried on odd-numbered *ESSA* satellites, *Nimbus-1* and

-2, ITOS and *NOAA-1.* Waveband: 0.45–0.65 μm. Spatial resolution: 2 km. Swath width: 830 km (2300 km from ESSA and ITOS satellites). The AVCS stored images on board (36 for ESSA satellites, 192 for Nimbus, ITOS and NOAA) for downloading to a receiving station.

AVHRR (Advanced Very High Resolution Radiometer) U.S. optical/infrared mechanically scanned imaging radiometer, carried on *TIROS-N* and *NOAA* satellites (NOAA-6 onwards). Wavebands (μm):

	1	2	3	4	5
TIROS-N	0.58–0.68	0.73–1.10	3.55–3.93	10.1–11.1	
NOAA-6, -8, -10	0.58–0.68	0.73–1.10	3.55–3.93	10.5–11.5	10.5–11.5
NOAA-7, -9, -11, -12, -14*	0.58–0.68	0.73–1.10	3.55–3.93	10.3–11.3	11.5–12.5

*The instrument carried by NOAA-7, -9, -11, -12 and -14 is designated AVHRR/2.

Spatial resolution: 1.1 km or 4 km at nadir. Swath width: 2600 km or 4000 km.
   HRPT (High Resolution Picture Transmission) data have 1 km resolution and are transmitted continuously. They cover a 2600 km swath (2048 pixels). **LAC** (Local Area Coverage) data also have 1 km resolution and are programmed for recording on board and downloading once per orbit (duty cycle 10%). **GAC** (Global Area Coverage) data are also stored on board for downloading once per orbit, but at a reduced spatial resolution of 4 km. **APT** (Automatic Picture Transmission) data have a spatial resolution of about 4 km and a swath width of 4000 km (2 bands only), and are transmitted continuously. Continuously transmitted data may be received from the S-band downlink and used freely.
   Future NOAA satellites will carry the AVHRR/3 instrument, which will include a band at 1.58–1.64 μm. It is also planned that this instrument will be carried on *Metop-1.*

URL: http://edcwww.cr.usgs.gov/glis/hyper/guide/avhrr

AVNIR (Advanced Visible and Near Infrared Radiometer) Japanese optical/near infrared *CCD* (pushbroom) imaging radiometer on *ADEOS* satellite. Wavebands: 0.42–0.50, 0.52–0.60, 0.61–0.69, 0.52–0.69, 0.76–0.89 μm. Spatial resolution: 16 m (8 m for 0.52–0.69 μm [panchromatic band]). Swath width: 80 km, selectable within ±40° of sub-satellite track.
   AVNIR-2, planned for inclusion on *ALOS*, will have a spatial resolution of 2.5 m (panchromatic), 10 m (multispectral), and a swath width of 35 km (panchromatic), 70 km (multispectral).

URL: http://www.eorc.nasda.go.jp/ADEOS/Project/Avnir.html

Azimuth direction The direction on the Earth's surface parallel to the motion of a *side-looking radar* or *synthetic aperture radar*, also called the along-track direction.

Azimuth shift The displacement of the image of a moving target in a *synthetic aperture radar* (SAR) image, relative to the position at which it would have been imaged if it had been stationary. The shift depends on the imaging geometry and the target velocity. It occurs as a result of the way SAR processing uses the Doppler shift of the backscattered radiation to determine the scatterer's azimuth coordinate; motion of the target introduces an extra Doppler shift which is interpreted as a displacement in the azimuth direction.

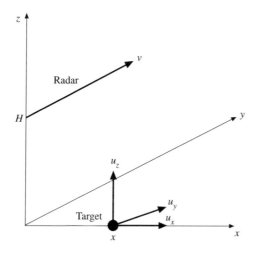

The diagram shows the geometry of azimuth shift in a Cartesian coordinate system that can be used to represent either a flat Earth geometry or a curved Earth surface. At some instant the target is located at $(x, 0, 0)$ and moves with velocity (u_x, u_y, u_z). The radar has velocity $(0, v, 0)$ and moves in the yz-plane with $z = H$. The azimuth (y) coordinate assigned to the target under these conditions is

$$\frac{Hu_z - xu_x}{v - u_y}.$$

Azimuth shift is apparent only if the target's motion is maintained during the SAR's coherence time, i.e. the time taken for the platform to move through the length of the synthetic aperture. If the motion is not maintained over this period, the target's image will be blurred. Azimuth shift is useful in some applications, for example determining ship velocities by comparing the moving ship with its stationary wake. However, it is a significant source of uncertainty in determining ocean wave spectra from SAR images.

Backscatter General term for the *scattering* of radiation through an angle of more than 90° from the forward direction. More specifically, scattering through an angle of 180°.

Backscatter coefficient A dimensionless quantity, represented by the symbol σ^0, denoting the effectiveness of a surface at *scattering* radiation incident upon it. It is normally used to describe the response of *radar* systems, and is defined by the equation

$$\sigma = \sigma^0 \, dA,$$

where σ is the backscatter cross-section of an element dA of the surface. σ is, in turn, defined as the area intercepting that amount of power which, if scattered isotropically, would give rise to the backscatter measured in the direction of observation. As a dimensionless quantity, the backscatter coefficient is often specified in *decibels*.

σ^0 is related to the *bidirectional reflectance distribution function R* by

$$\sigma^0 = 4\pi R \cos\theta_0 \cos\theta_1,$$

where θ_0 and θ_1 are the angles between the incident and scattered radiation, respectively, and the surface normal. The backscatter coefficient is sometimes also written as

$$\gamma = \frac{\sigma^0}{\cos\theta_0}.$$

The backscatter coefficient is dependent on the frequency, observation geometry and polarisation states of the incident and scattered radiation as well as on the properties of the scattering surface. For a given frequency and geometry, it is common to express the backscatter coefficient as

$$\sigma^0_{pq},$$

where p denotes the polarisation state of the incident radiation and q that of the scattered radiation.

Backscatter lidar See *lidar*.

Backscatter Ultraviolet Spectrometer See *BUV*.

Balkan-1 Russian laser profiler carried on *Mir-1*. Wavelength: 532 nm (Nd-YAG laser). Spatial resolution: 150 m (horizontal), 3 m (vertical). Pulse repetition frequency: 0.18 Hz.

Balkan-2 Russian laser profiler/backscatter *lidar* planned for inclusion on *Almaz-1B*. Wavelength: 532 nm (Nd-YAG laser). Spatial resolution: 16 km (horizontal), 10 m (vertical, as lidar), 1 m (vertical, as laser profiler). Swath width: Steerable ±70 km from nadir. Pulse repetition frequency: 1 Hz.

Banding A periodic error in image brightness that can be caused by *scanning systems* that use more than one detector to acquire the image. Banding can be particularly problematic in *Landsat MSS* images, since the MSS scanning system used six different detectors to image adjacent scan lines. The same detector thus viewed only every sixth scan line, so that any uncorrected *calibration* errors resulted in a spurious six-pixel periodicity in the along-track direction of the image. Banding, also called **striping**, can be at least partially removed by *destriping* algorithms.

Band interleaved by line See *image format*.

Band interleaved by pixel See *image format*.

Band sequential format See *image format*.

Baseline The separation between two positions of a sensor from which slightly different images of the same area are obtained. The differences contain information on the topography of the area. See *interferometric SAR, stereophotography*.

Bathymetry Measurement of the depth of a body of water. Bathymetry from satellite remote sensing data is possible in some limited circumstances. In very clear water, depths up to about 10 m can be inferred from visible-wavelength imagery, through the effect of absorption by the water column on the light reflected from the bottom, or from *synthetic aperture radar* imagery, through the effects of bottom topography on the refraction and diffraction of surface waves. Limited information on deep-ocean bathymetry can be obtained from *radar altimeter* observations, since the topography of the ocean surface follows the *geoid* (provided that there are no ocean currents and that the effects of tides, waves and atmospheric pressure variations have been removed), and variations in the geoid height reflect variations in the bottom topography.

Beam-limited Term used to describe the operation of a *radar altimeter* when the effective spatial resolution is determined by the *power pattern* of the antenna rather than by the duration of the (compressed) pulse. See *pulse-limited*.

Beamwidth The effective angular width of the *power pattern* of an *antenna*, measured in a given plane. In practice, the effective beamwidth is usually

23

defined as the width of the beam at half the peak value and referred to as the half-power beamwidth. The beamwidth is usually specified in each of two principal planes, the azimuth (horizontal) plane and the elevation plane. If the direction of the beam is denoted by the z-axis, the horizontal direction by the y-axis and the vertical direction by the x-axis, the beamwidth in the xz-plane is denoted by β_{xz} and that in the yz-plane by β_{yz}. If the antenna dimensions are D_x and D_y along x and y respectively, then

$$\beta_{xz} = \frac{k_x \lambda}{D_x}, \qquad \beta_{yz} = \frac{k_y \lambda}{D_y},$$

where k_x and k_y are illumination factors that characterise the electric field distribution across the antenna aperture, λ is the wavelength, and β is measured in radians. For a uniformly illuminated aperture, $k_x = k_y = 0.88$, which yields the narrowest possible beamwidths. Tapered illumination is often used to reduce the side-lobe levels of the power pattern, which leads to larger values of k_x and k_y. For steep tapers, with correspondingly very low side-lobe levels, k_x and k_y may be as large as 2.

Bhaskara-1, -2 The first experimental remote-sensing satellites launched by India, in 1979 and 1981 respectively. Principal instruments: vidicon (one optical and one near infrared channel), three-channel passive microwave radiometer.

Bhattacharyya distance See *separability*.

Bidirectional reflectance distribution function A function characterising the amount of electromagnetic radiation reflected (scattered) from a surface, as a function of the directions of the incident and reflected radiation. The bidirectional reflectance distribution function (BRDF) is defined as the ratio of the reflected *radiance* to the incident *irradiance*. It is normally a function of the directions of both the incident radiation and the reflected radiation, and is denoted by

$$R(\theta_0, \phi_0, \theta_1, \phi_1),$$

where θ_0 and θ_1 are the angles between the incident and reflected radiation, respectively, and the surface normal, and ϕ_0 and ϕ_1 are the corresponding azimuth angles (see figure). The BRDF is symmetric with respect to the incident and reflected directions, i.e.

$$R(\theta_0, \phi_0, \theta_1, \phi_1) = R(\theta_1, \phi_1, \theta_0, \phi_0).$$

The **reflectivity** of the surface is a function only of the incidence direction, and is defined as the ratio of the *radiant exitance* to the irradiance. It is given in terms of the BRDF by

$$r(\theta_0, \phi_0) = \int\limits_{\theta_1 = 0}^{\pi/2} \int\limits_{\phi_1 = 0}^{2\pi} R(\theta_0, \phi_0, \theta_1, \phi_1) \cos \theta_1 \sin \theta_1 \, d\theta_1 \, d\phi_1.$$

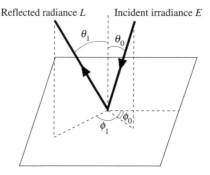

Reflected radiance L Incident irradiance E

The **diffuse albedo** (sometimes called the hemispherical albedo) is the ratio of the total scattered power to the total incident power when the latter is distributed isotropically, and is given by

$$r_d = \frac{1}{\pi} \int\limits_{\theta_0=0}^{\pi/2} \int\limits_{\phi_0=0}^{2\pi} r(\theta_0, \phi_0) \cos\theta_0 \sin\theta_0 \, d\theta_0 \, d\phi_0.$$

See also *albedo, backscatter coefficient, bidirectional reflectance factor, Lambertian reflection, specular reflection.*

Bidirectional reflectance factor The bidirectional reflectance factor (BRF) is defined as the ratio of the flux scattered into a given direction by a surface under given illumination conditions, to the flux scattered in the same direction by a perfect *Lambertian* scatterer under the same conditions. The BRF $f(\theta_0, \phi_0, \theta_1, \phi_1)$ is related to the *bidirectional reflectance distribution function* by

$$f(\theta_0, \phi_0, \theta_1, \phi_1) = \frac{R(\theta_0, \phi_0, \theta_1, \phi_1)}{\pi}.$$

Calibration standards can be manufactured to have a BRF that is uniform over a wide range of wavelengths and scattering angles.

BIL See *image format.*

Bilinear interpolation See *resampling.*

Biological productivity, oceans The primary biological productivity in oceans is the 'fixing' of carbon (conversion from dissolved CO_2 to biological carbon) by *phytoplankton*. Detection and quantification of biological productivity, for estimation of its contribution to the global carbon cycle and for the identification of potential fishing areas, can be performed from satellite sensor observations of *ocean colour.*

Biomass The total mass of dry organic matter present per unit area of the Earth's surface. The above-ground biomass of terrestrial vegetation is commonly estimated using *vegetation indices.*

BIP See *image format.*

Bistatic See *radar equation.*

Black body A body that absorbs all the radiation incident upon it, reflecting none. Such a body also has an *emissivity* of 1. A black body at an absolute temperature T emits radiation whose properties are characterised only by the value of T (see *Planck distribution*). This radiation is called black-body radiation.

BNSC (British National Space Centre) See *CEOS.*

Bouguer's law See *Lambert–Bouguer law.*

Boundary layer, atmospheric The lowest region of the *atmosphere*, of the order of 1 km thick, in which transport processes are dominated by wind turbulence and by convection. The majority of molecular and particulate species (particularly *water vapour* and *aerosols*) found in the boundary layer are generated by surface interactions.

Box classifier See *supervised classification.*

Bragg scattering A surface *scattering model* based on coherent scattering from periodic structure in the surface. Bragg scattering is particularly important in microwave remote sensing of ocean surfaces, where backscatter is dominated by coherent scattering from a specific wave component of the ocean surface with a wavelength Λ equal to $\lambda/(2\sin\theta)$, where λ is the wavelength of the incident radiation and θ is the *incidence angle*. The figure shows that this is equivalent to requiring that twice the path difference, $2\Lambda\sin\theta$, should be equal to one incident wavelength. The factor of two arises because the wave must travel over the same path twice.

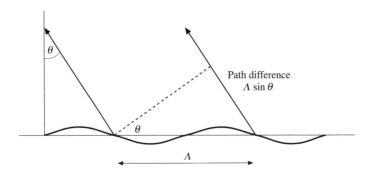

BRDF See *bidirectional reflectance distribution function.*

Brewster angle If electromagnetic radiation propagating *in vacuo* is incident on a loss-free medium having a planar surface, the *Fresnel coefficient* for *parallel-*

polarised radiation is zero when the *incidence angle* is equal to the Brewster angle θ_B. If the medium has refractive index n, the Brewster angle is given by

$$\tan\theta_B = n.$$

BRF See *bidirectional reflectance factor*.

Brightness

1. See *IHS display*.
2. See *tasselled-cap transformation*.

Brightness temperature The brightness temperature of a body that is emitting thermally generated radiation is the temperature that the body would need to have if it were a *black body* emitting the same amount of radiation. For a body of *emissivity* ε at absolute temperature T, the brightness temperature T_b at wavelength λ satisfies the equation

$$\varepsilon(\exp[hc/\lambda k T_b] - 1) = \exp[hc/\lambda k T] - 1,$$

where h is Planck's constant, c is the speed of light *in vacuo* and k is Boltzmann's constant. If the frequency is low enough that the *Rayleigh–Jeans approximation* is valid, this equation simplifies to

$$T_b = \varepsilon T.$$

The brightness temperature of radiation is defined in terms of its *radiance* through the *Planck distribution*.

British National Space Centre The British space agency. See *CEOS*.

URL: http://www.open.gov.uk/bnsc/bnschome.htm

BSQ See *image format*.

BUFS-4 Russian ultraviolet/near infrared spectrometer for ozone profiling, planned for inclusion on *Meteor-3M* satellite series. <u>Wavebands</u>: 0.312–0.349 μm, 0.738–0.762 μm. <u>Spatial resolution</u>: $3° \times 3°$.

Building materials, electromagnetic properties Typical *emissivities* in the wavelength range 8–14 μm are listed below:

Material	Emissivity
aluminium, polished	0.06
asphalt	0.96
brick	0.93–0.95
concrete	0.92–0.97
glass	0.94
paint	0.94
plaster	0.91
wood	0.90–0.95

BUV (Backscatter Ultraviolet Spectrometer) U.S. nadir-viewing ultraviolet spectrometer, carried on *Nimbus-4* satellite. Wavebands: 12 bands between 0.25 and 0.34 μm.

The purpose of BUV was to provide vertical profiles of atmospheric ozone concentrations. It was the forerunner of the *SBUV* instrument.

Calibration

1. The process of relating the measured output (spatial, radiometric, spectral, polarimetric etc.) of a sensor to the true value of the quantity.
2. The results of the process (1).

Canadian Space Agency See *CSA*.

Candela The unit of luminous intensity, equal to one lumen per steradian. See *photometric quantities*.

Canonical components The canonical components of a multi-band image are linear combinations of the bands, chosen in such a way that classes (clusters) of data are maximally separable using the first canonical component. The degree of *separability* decreases from the first canonical component to the second, from the second to the third, and so on.

Canonical component analysis is similar to *principal component* analysis (PCA), but, unlike PCA, takes account of the *clustering* of the data in the N-dimensional feature space (e.g. the N spectral bands of a multispectral image). If there are M clusters, such that the ith cluster contains n_i vectors in feature space, the mean value of these vectors is $\mathbf{m}_i$ and their *covariance matrix* is $\mathbf{C}_i$, the among-class covariance matrix $\mathbf{C}_A$ is given by

$$\mathbf{C}_A = \frac{1}{M} \sum_{i=1}^{M} ((\mathbf{m}_i - \mathbf{m}_0)^T (\mathbf{m}_i - \mathbf{m}_0)),$$

where

$$\mathbf{m}_0 = \frac{\sum_{i=1}^{M} n_i \mathbf{m}_i}{\sum_{i=1}^{M} n_i}$$

and the superscript T denotes a vector transpose, and the within-class covariance matrix $\mathbf{C}_W$ is given by

$$\mathbf{C}_W = \frac{\sum_{i=1}^{M} (n_i - 1)\mathbf{C}_i}{\sum_{i=1}^{M} n_i}.$$

The jth canonical component can be written as a vector $\mathbf{c}_j$ in feature space, representing a linear combination of $x_1, x_2, \ldots, x_N$. It satisfies the generalised eigenvalue equation

$$(\mathbf{C}_A - \lambda_j \mathbf{C}_W)\mathbf{c}_j = 0$$

and a normalisation condition

$$\mathbf{c}_j^T \mathbf{C}_W \mathbf{c}_j = 1.$$

The order of the canonical components is such that $\lambda_1 > \lambda_2 > \ldots$

Carbon dioxide An important, spatially and temporally variable, radiatively active constituent of the Earth's *atmosphere*. Chemical formula CO_2. Profiling of atmospheric carbon dioxide can be carried out using the infrared absorption lines. The most important of these have wavelengths of 1.95, 2.0, 2.1, 2.7, 2.8, 3.3, 4.3 and 15 μm.

Carterra-1 (Also known as CRSS, SIS) High-resolution optical/near infrared imager, operating in panchromatic and multispectral modes, planned for inclusion on *Ikonos-1*. Wavebands: 0.45–0.90 μm (panchromatic); 0.45–0.52, 0.50–0.60, 0.63–0.69, 0.76–0.90 μm (multispectral). Spatial resolution: 4 m (multispectral); 1 m (panchromatic). Swath width: 11 km.

URL: http://www.spaceimage.com/home/products/index.html

Case 1/Case 2 waters See *ocean colour*.

CAST (Chinese Academy of Space Technology) See *CEOS*.

C band Subdivision of the microwave region of the *electromagnetic spectrum*, covering the frequency range 3.9–6.2 GHz (wavelengths 48–77 mm).

CBERS (China–Brazil Earth Resources Satellite) International satellite with a Brazilian payload carried on a Chinese platform, known as Ziyuan-1 in China. Operated by CAST (China) and INPE (Brazil) and scheduled for launch in the late 1990s. Objectives: Global (especially Brazil) land, ocean, atmosphere observations. Orbit: Circular *Sun-synchronous LEO* at 778 km altitude. Period 100.3 minutes; inclination 98.5°; equator crossing time 10:30 (descending node). *Exactly repeating orbit* (373 orbits in 26 days). Principal instruments: *HRCC, IR-MSS, WFI*. CBERS will also carry a data-collection package similar to *Argos*.

CCD (Charge-coupled device) An imaging *electro-optical sensor*. A CCD can record radiation from a ground resolution element (see *rezel*) for representation within a *pixel* in an image. The simplest CCD array is linear and this can be used to record radiation from a line of ground resolution elements in 'imaging mode' (e.g. the *SPOT HRV* sensors), or from one ground resolution element in 'spectral mode' (e.g. the Spectron radiometer). In its imaging mode, line after line of radiation can be built up to form an image as the aircraft or satellite moves forward. In comparison with a scanner the linear CCD array has more sensors to calibrate and is restricted in the range of wavelengths it can sense. However, it possesses many advantages over the scanner: lighter weight; smaller size; lower power requirement; no moving parts; longer life; greater reliability; greater geometric accuracy; greater radiometric accuracy; lower cost; longer dwell time on each ground resolution element. The last advantage assures a larger signal and the potential to increase the signal-to-noise ratio, the spatial resolution or the spectral resolution.

Today many optical sensors use one or several one-dimensional or two-dimensional CCD arrays. These arrays have made possible improvements in image data quality and the development of fine spatial resolution sensors, *imaging radiometers* and high-performance spectroradiometers. See also *step-stare imager*.

CCT (Computer-compatible tape) Magnetic tape used for storage of image (and other) data; now largely superseded by 8 mm cartridge (exabyte) types and compact disc storage.

Centre National d'Etudes Spatial See *CNES*.

CEOS (Committee on Earth Observation Satellites) International body established in 1984 to coordinate international activity in spaceborne remote sensing. The member organisations of CEOS are ASI (Agenzia Spaziale Italiana, Italy), BNSC (British National Space Centre), CAST (Chinese Academy of Space Technology), *CNES* (France), *CSA* (Canada), CSIRO (Commonwealth Scientific and Industrial Research Organisation, Australia), *DARA* (Germany), *ESA* (Europe), EUMETSAT (European Organisation for the Exploitation of Meteorological Satellites), the European Union, *INPE* (Brazil), *ISRO* (India), *NASA* (U.S.A), *NASDA* (Japan), *NOAA* (U.S.A.), NRSCC (National Remote Sensing Centre of China), NSAU (National Space Agency of Ukraine), ROSGYDROMET (Russian Federal Service for Hydrometeorology and Environmental Monitoring), RSA (Russian Space Agency), SNSB (Swedish National Space Board) and STA (Science and Technology Agency, Japan).

URL: http://ceos.esrin.esa.it/

See also *legal and international aspects*.

CERES (Clouds and Earth's Radiant Energy System) U.S. optical/infrared scanning radiometer, carried on *TRMM* and planned for inclusion on *EOS-AM* and *EOS-PM* satellites. Wavebands: 0.3–50, 0.3–5.0, 8–12 μm. Spatial resolution: 21 km at nadir. Swath width: Limb to limb.

CERES is intended for Earth radiation budget and cloud studies, and is based on the *ERBE* instrument.

URL: http://asd-www.larc.nasa.gov/ceres/ASDceres.html

Charge-coupled device See *CCD*.

Chemistry, atmospheric Trace gases in the Earth's *atmosphere* can be significant through being radiatively active ('greenhouse gases', which are directly involved in the Earth's *climate*), chemically active (in which case they affect the Earth's environment), or through modification of the global *ozone* cycle (in which case they have both climatological and environmental effects). These trace gases can be detected and measured by the amount of electromagnetic radiation that they absorb, emit or scatter. *Atmospheric sounding* of most trace gases requires high spectral resolution in order to resolve individual absorption or emission lines, although lower spectral resolution is required if only column-integrated values are required. Lower spectral resolution is also needed for profiling of ozone concentrations, since ozone has particularly strong absorption lines. Profiling of concentrations in the *troposphere* is more difficult than in the *stratosphere*. Currently, most of the important instruments for atmospheric chemistry measurements are carried by the *UARS* satellite. In the near future, instruments such as *GOME, ILAS, IMG, MIPAS, Sciamachy* and *TES* will enhance the ability to measure atmospheric trace gases.

China–Brazil Earth Resources Satellite See *CBERS*.

Chinese Academy of Space Technology The Chinese space agency. See *CEOS*.

Chlorophyll The most important photosynthetic pigment in plants (see *leaf*). Chlorophyll-a has absorption maxima at $0.43\,\mu m$ (blue) and $0.66\,\mu m$ (red); chlorophyll-b at $0.45\,\mu m$ and $0.65\,\mu m$.

Chlorosis See *geology*.

Circular polarisation See *polarisation*.

CLAES (Cryogenic Limb Array Etalon Spectrometer) U.S. infrared limb-sounding spectrometer, carried on *UARS*. Wavebands: 3.5, 6.0, 8.0, $12.7\,\mu m$. Spatial resolution: 2.8 km (vertical); 480 km (horizontal). Altitude range: 10–60 km in 20 steps.

CLAES was used to deduce concentrations of various atmospheric gases (CO_2, H_2O, CH_4, O_3, and several nitrogen and chlorine compounds) from measurements of thermal emission. It ceased operation in May 1993.

URL: http://www.lmsal.com/9120/CLAES/claes_homepage.html

Clark U.S. satellite, operated by NASA and planned for launch in 1998. Objectives: Proof-of-concept mission for small, low-cost spacecraft. Orbit: Circular *Sun-synchronous* LEO at 496 km altitude; inclination 97.3°, period 92.3 minutes. Principal instruments: *Earthwatch imager*, *MicroMAPS*. The satellite will also carry a solar–terrestrial physics package and a laser retroreflector for ground-based *lidar* measurements of the atmosphere.

URL: http://www.crsp.ssc.nasa.gov/ssti/clark/clark.htm

Classification, image The aspect of *image processing* in which quantitative decisions are made on the basis of the data present in the image, grouping *pixels* or regions of the image into classes representing different ground-cover types. The output of the classification stage may be regarded as a thematic map rather than an image. Its accuracy is normally assessed using an *error matrix*.

Classification techniques can be broadly divided into two types: *supervised classification* and *unsupervised classification*. In supervised classification, information about the distribution of ground-cover types in part or parts of the image is used to initiate the process. Pixels or groups of pixels corresponding to known cover types are called *training data* or training areas, and are used to 'train' the classification process to recognise other, similar, pixels. In unsupervised classification, the entire image is first analysed by *clustering* to find distinguishable classes of pixels present within it. After this stage has been completed, the classes present within the image are associated with classes present on the ground by comparison with training data. In essence, therefore, supervised classification forces the image classification to correspond to user-defined ground-cover classes, but does not guarantee that the classes will be separable; whereas unsupervised classification forces the classes to be separable but does not guarantee that they will correspond to the ground-cover classes required by the user. *Hybrid classification* algorithms combine features of both approaches.

Climate The climate of a location is the synthesis of the day-to-day values of the main meteorological elements that affect the site. Factors that affect climate include *precipitation*, temperature, cloud cover, *wind speed* and direction, sunshine and *humidity*. Satellite remote sensing systems can provide data on many of these quantities and the data are a valuable supplement to the *in situ* observations, which tend to be distributed very unevenly around the globe with a bias towards the heavily populated areas of Europe and North America. Satellite data also have the advantage of being collected on a global basis by a single sensor and the data can be processed in a consistent way for all locations. Some satellite data sets, such as the global *cloud statistics*, *sea ice* coverage and global *sea surface temperature* data sets, are very important for climate studies and are maturing into reliable products, although the data sets are of short duration compared to *in situ* observations. Other data, such as climatological fields of precipitation over the ocean, are in an early stage of development, and there is still a great deal of development work taking place on the processing algorithms. Some quantities, such as surface heat and moisture flux and near surface air temperature, cannot be derived in a reliable manner from observations from satellite sensors at present.

Cloud classification *Clouds* form in the atmosphere as a result of many different physical processes. They are visible aggregations of water droplets or ice crystals with a base above ground level. The nature of the atmospheric structure, the physical form of the cloud, including the shape of individual cloud elements, and the phase of the water (liquid or solid) which makes up the cloud provide the most common basis for cloud classification. Methods of classifying clouds on the basis of height of formation and shape are used by surface observers (e.g. cirrus, cirrostratus, cirrocumulus, altostratus, nimbostratus, altocumulus, cumulus, stratocumulus, stratus and cumulonimbus) but are less suited to classification of clouds from satellite-based radiance measurements because many of the criteria for these classifications are not available from satellite information. In the case of high-resolution satellite imagery, human classification of cloud types may be possible. In most types of imagery, the resolution is insufficient to permit the clear identification of such cloud types. Automatic classification of clouds in remote sensing can only be on the basis of characteristics which can be extracted from the radiance field measured by the satellite. The classification generally follows the *cloud detection* process.

The simplest form of automated classification from satellite imagery is therefore based on the brightness of the clouds at visible wavelengths and the temperature of the cloud as indicated by the brightness at thermal infrared wavelengths. Two radiance channels are generally considered to be the minimum requirement. The detail of classification from remote sensing imagery is limited by the complexity of the algorithms used. More detailed information on cloud character can be obtained by examining the spatial and temporal variance of the radiance field. The altitude of cloud-tops can be determined directly by *laser profiling*, and the water content by *lidar* or *passive microwave radiometry*. A very simple cloud classification, due to the ISCCP (International Satellite Cloud Climatology Project), based on the cloud's optical thickness (see *Lambert–Bouguer* law) and the cloud-top pressure, is shown in the figure.

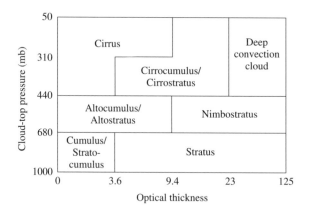

Cloud detection Detection of clouds in satellite imagery is straightforward in most circumstances (see also *cloud masking*). In most situations, clouds are brighter than the background surface at visible wavelengths and colder than the background surface at thermal infrared wavelengths. Infrared imagery is usually viewed as a negative image so that clouds appear white. Several problem areas for cloud detection exist, and cloud detection is difficult if the image pixel size is significantly larger than the size of individual cloud elements. This is particularly problematic for thin cirrus, low stratiform cloud decks, polar regions, and multilayer cloud systems. Cloud detection is difficult at night in areas where the thermal contrast between the surface and the cloud is low, for example in the case of marine stratocumulus regions and areas of fog. In regions of thin cirrus, the observed radiance is a combination of the radiances from the cloud and the ground. In the polar regions and other areas of snow cover, the contrast is low at visible wavelengths and the thermal gradient in the atmosphere is often weak. In the polar winter, where a strong surface inversion forms, the thermal contrast is inverted.

Detection of clouds in problem areas can be enhanced by the use of multispectral imagery to detect the signatures of cloud types. Predominantly, cloud detection relies on optical and thermal infrared observations, but wavelength regions in the near infrared have been used to distinguish between clouds and snow (1.6 μm) and to detect the presence of fog (3.7 μm). Such multispectral analysis offers the opportunity to develop cloud type signatures for improved classification of clouds. Pattern recognition techniques have also been applied to problem areas such as wintertime polar clouds.

Cloud masking Cloud masking is essentially the same process as *cloud detection*, except that the focus is on the elimination of cloud-affected pixels from further analysis. When determining surface parameters such as surface temperature and vegetation indices from satellite radiance data, the presence of clouds can seriously contaminate the results.

A number of techniques have been used for cloud masking. A simple threshold *brightness temperature* (e.g. *AVHRR* channel 5) can be used to indicate cloud-affected pixels. Alternatively, a ratio $Q = R_1/R_2$, where R_1 is the optical-band reflectance and R_2 the near infrared reflectance, can be used to distinguish between ocean (Q generally less than 1), land (Q greater than 1) and cloud ($Q \approx 1$). Some techniques have used the difference in brightness temperature between AVHRR channels 4 and 5 as indicators of the presence of cloud. The spatial coherence of cloud-affected areas is generally larger than that of cloud-free areas, and this can also be used for cloud masking although the technique is not useful where the background variance is large, e.g. in coastal regions. Most cloud-masking schemes involve the derivation of empirical thresholds and their success depends on the geographical area of application.

Clouds Clouds are collections of ice crystals, water droplets or a mixture of the two that have their base above the surface of the Earth. They form when air is cooled sufficiently so that its temperature falls below the dew point, i.e. the temperature at which it becomes saturated with respect to water. Under these conditions the excess water condenses as water droplets or ice crystals. The cooling of the air usually occurs because of expansion as it rises through the atmosphere. This can take place for a number of reasons and affects the appearance of the clouds. Large-scale atmospheric motion, such as at a frontal surface, usually gives layer cloud, while convection above a relatively warm surface often results in more isolated clouds. Forced ascent of air over high ground is also an important mechanism in the formation of cloud. The water droplets making up a cloud are often supercooled, but they must have a diameter of less than 200 μm otherwise they are classed as rain or drizzle. The vast majority of clouds are found below the *tropopause*, which is at a height of about 11 km in mid-latitudes, although some important cloud types occur in the *stratosphere*.

Although the cloud-climate feedback is insufficiently understood, it is clear that clouds play an important part in determining the *Earth radiation budget*. There are two main effects. At optical wavelengths they act to reduce the amount of heating by reflecting radiation back to space. At thermal infrared wavelengths, they absorb radiation emitted from the Earth's surface and re-emit it back to the surface and into space (at a lower temperature). Since clouds are collections of small refracting particles, they are effective scatterers of solar radiation. This scattering is well described by the *Mie* theory. The *albedo* of clouds is dependent on a number of factors including their thickness and droplet size but they typically have values in the range 50% to 65%.

Cloud properties are generally discussed in terms of their liquid water path, optical depth or liquid water content. At thermal infrared wavelengths, clouds (with the exception of thin cirrus) behave as *black bodies*. Most analyses of cloud radiation interactions assume that the clouds are plane parallel and that the effects of the sides can be ignored. In many environments, where broken cloud fields are common, the contribution of radiation reflected and radiated from cloud sides must be considered. Cloud sides act to enhance both infrared and optical-band radiation with respect to the linear combination of clear and overcast conditions.

Most analyses of the effects of clouds are performed in terms of the concept of cloud radiative forcing. In a region large enough to consist of both clear and cloud-covered areas (e.g. a latitude zone, ocean basin or the whole globe) the net radiative heating H for the surface–atmosphere column is

$$H = S(1 - \alpha) - F,$$

where S is the solar irradiance, α is the albedo and F is the emitted thermal infrared flux. The cloud forcing is then $H - H_{clr}$, where H_{clr} is the net heating under clear sky conditions. H_{clr} can be written as the sum of short-wave and long-wave forcings, $C_{sw} + C_{lw}$. The cooling effect (negative cloud forcing) in the short-wave region is comparable to, but slightly larger than, the warming effect (positive cloud forcing) in the long-wave region. Results from one study give $C_{sw} = -44.5\,\mathrm{W\,m^{-2}}$ and $C_{lw} = +31.3\,\mathrm{W\,m^{-2}}$.

Clouds and Earth's Radiant Energy System See *CERES*.

Cloud statistics Prior to observations of clouds from satellite sensors, a range of climatologies describing cloud distribution, derived from surface observations, were available. Since low clouds tend to obscure high level clouds from a surface observer and high clouds obscure low clouds when viewed from a satellite, the two observations rarely agree on the character of cloud cover over the globe. The range of viewing angles which make up the surface observer's view of the sky and the range of nadir angles which are sampled in a single satellite image result in very different sets of statistics from the two methods. Differences in satellite frequency response and in the nature of the algorithm used in the *cloud detection* and *cloud classification* stages mean that agreement between different satellite climatologies is very variable.

The International Satellite Cloud Climatology Project (ISCCP) was designed to provide a global archive of satellite-derived cloudiness information. It relies primarily on variable thresholds determined from statistical analyses of radiances retrieved for individual locations. The global mean ISCCP cloud amount for 1984–1988 was 62.6%. The mean annual cloud amount is greater over the ocean than over the land. Seasonal variation in global mean cloud amount is very small (less than 1% from month to month). Although the differences between mean ISCCP cloud amounts and other cloud climatologies are usually small (about 6%), they can be as high as 14% in the polar regions. As well as deriving statistics for the distribution of cloud amount, some attempts have been made to characterise cloud cover using higher-order statistics, for example the β distribution, the Burger distribution and fractal analysis.

Cloud temperature The tops of cloud layers are, in the case of thick clouds, the emitting surfaces from which infrared radiation is received by satellite radiometers. The temperature of the top of the cloud therefore affects the amount of radiation received at satellite level. This quantity allows the height of the cloud top to be determined, provided that the profile of temperature through the depth of the atmosphere is known. Cloud top temperatures can range from 10 or 20 °C to −60 °C or colder and are essentially a function of the height of the cloud in the atmosphere. Low cloud is found in approximately the lowest 2 kilometres of the atmosphere and cloud top temperatures range from close to the surface temperature to about 0 °C when we consider the mean mid-latitude atmospheric profile. Medium level cloud is found between about 2 and 6 kilometres above the surface, which in mid-latitudes has cloud top temperatures between about 0 °C and −25 °C. Although these temperatures are below the freezing point of water, the clouds are usually composed of supercooled water droplets. High level cloud, between about 6 and 13 kilometres above the surface, is often composed of ice crystals and the temperatures observed range from about −25 °C to −60 °C.

Cloud top temperatures can be combined with measurements of cloud thickness to estimate *precipitation*.

Cloud water content Cloud water content is usually expressed as the amount (in grams per cubic metre) of water, in the liquid or solid state, that is contained within a cloud. The median values of water content for low level layer cloud (e.g. stratus) in mid-latitudes are about $0.2\,\mathrm{g\,m}^{-3}$ and for medium level clouds (e.g. altostratus) about $0.1\,\mathrm{g\,m}^{-3}$. However, the variability of water content is large and *in situ* measurements have indicated values of up to five times these figures. For convective cloud, water content can be much higher with values of around $2\,\mathrm{g\,m}^{-3}$ being measured near the top of stratocumulus and cumulus. The water content of individual clouds is dependent on a number of factors, including cloud base temperature and the degree of vertical development of the cloud.

Clustering The process of identifying groups of *pixels* in an image that have similar properties. Commonly the first step in an *unsupervised classification* of an image.

The properties (features) of a pixel can be specified by a vector $\mathbf{x}$ in N-dimensional feature space. The components of this vector will often be the digital numbers or reflectances in each of N spectral bands, but could also be, for example, radar backscatter coefficients in different polarisation states, texture parameters, or single-band digital numbers in a multi-date composite image.

The basis of all clustering algorithms is a measure of the similarity of two pixels with vectors $\mathbf{x}_1$ and $\mathbf{x}_2$. The simplest of these is the **Euclidean distance**

$$|\mathbf{x}_1 - \mathbf{x}_2|^{1/2},$$

although the square of this quantity is often used to reduce the computational requirement. Another measure of similarity, slightly easier to calculate, is given by

$$\sum_{i=1}^{N} |x_{1i} - x_{2i}|,$$

where x_{1i} is the ith component of the vector $\mathbf{x}_1$, and similarly for $\mathbf{x}_2$.

Clustering of the pixels in an image is usually performed iteratively, using the *isodata* algorithm. Other methods include hierarchical clustering and single-pass clustering. Once the clusters have been defined, they are often modified by splitting, merging or deleting clusters. Splitting involves breaking into two or more clusters a single cluster that is excessively elongated in feature space, or that shows evidence of bi- or multi-modality. Merging is the combining of clusters that show insufficient *separability*. Clusters that contain too few pixels (typically less than $10N$) for further analysis are deleted.

CNES Centre National d'Etudes Spatiale, the French Space Agency.

URL: http://www.cnes.fr/

Coastal Zone Color Scanner See *CZCS*.

Coherence

1. The property possessed by a system of waves (usually electromagnetic radiation) when there is a fixed phase difference between the signal measured at the same location but at different times (temporal coherence) or between the signal measured at the same time but at different locations (spatial coherence).
2. The availability of both amplitude and phase information from a detected signal, for example in *synthetic aperture radar* systems.

Columbus Polar Platform See *Envisat*.

Committee on Earth Observation Satellites
See *CEOS*.

Commonwealth Scientific and Industrial Research Organisation
The Australian organisation with responsibility for space research. See *CEOS*.

Compression of data Data compression techniques are important in remote sensing because of the large volumes of data involved (for example, one *Landsat TM* image requires over 200 megabytes). Compression methods are either reversible, in which case all the information in the image is recoverable, or irreversible, in which case there is some loss of information.

The simplest **irreversible (lossy)** compression methods involve cropping images to remove uninteresting areas, or sub-sampling them to reduce their spatial resolution and size at the same time. Similarly, the *Fourier transform* or *Hadamard transform* of an image can be formed, and truncated to remove the higher spatial frequencies. Lossy compression is increasingly being used for the transmission of image data, particularly using the TIFF, GIF and JPEG formats.

Reversible (lossless) data compression methods can be divided into two types: those that operate on a pixel-by-pixel basis, recoding the sequence of binary digits needed to specify a pixel's value so that commonly occurring values are assigned short sequences, and those that make use of spatial uniformity in an image. *Huffman coding* is a common example of the former; *run-length encoding* and *tesseral addressing* are examples of the latter, although tesseral addressing can also be used as an irreversible method. Lossless compression is widely used for storage and transmission of image data.

Computer-compatible tape See *CCT*.

Confusion matrix See *error matrix*.

Consumer's accuracy See *error matrix*.

Contextual classification Modification of an image *classification* procedure to take into account the likelihood that neighbouring pixels should be assigned

to the same class, for example by incorporating into the discriminant function of a *supervised classification* a 'cost function' that penalises pixel-to-pixel variation in classification.

Contrast enhancement Radiometric transformation of an image to improve its visual interpretability. The operation can be specified by a transfer function f such that

$$I_d(p) = f(I_i(p)),$$

where $I_i(p)$ is the *digital number* of pixel p in the image, and $I_d(p)$ is the digital number used to represent the pixel in the display. The same transfer function f is used for all pixels in the image, or a specified part of it. The transfer function is chosen to maximise the use made of the radiometric resolution of the display unit, by expanding the width of the image *histogram.*

The simplest contrast enhancement is a **linear enhancement** or **linear stretch**. The transfer function has the form

$$f(x) = ax + b$$

where a and b are constants and $a > 1$. If the image, or a feature of interest within it, is represented by the range of digital numbers from I_{min} to I_{max}, the display will be represented by the range from $aI_{min} + b$ to $aI_{max} + b$. These values can be chosen to cover the entire range of available digital numbers, or even to exceed it at the lower or upper end of the display range (saturation of the display). Such saturation may be desirable from the point of view of increased radiometric separation of features of interest, if the pixels about which information is lost (through saturation) are not interesting.

Many image processing systems can provide **automatic linear enhancement**. The values of a and b are chosen on the basis of the statistical properties of $I_i(p)$, typically as

$$a = \frac{I_{d\,max} - I_{d\,min}}{N\sigma},$$

$$b = I_{d\,max}\left(\frac{1}{2} - \frac{\mu}{N\sigma}\right) + I_{d\,min}\left(\frac{1}{2} + \frac{\mu}{N\sigma}\right),$$

where μ and σ are, respectively, the mean and standard deviation of the values of $I_i(p)$, $I_{d\,min}$ and $I_{d\,max}$ are, respectively, the minimum and maximum values of $I_d(p)$ that can be displayed (normally 0 and $2^n - 1$ for an n-bit display), and N is usually set to 6. This transformation has the property that values of $I_i(p)$ less than $\mu - N\sigma/2$ or greater than $\mu + N\sigma/2$ will lead to saturation of the display.

Various forms of **non-linear contrast enhancement** are in common use. The simplest of these is the two- (or multi-) part linear enhancement, in which the transfer function f consists of a number of linear enhancements, with different values of a and b applying over different ranges of $I_i(p)$. Other non-linear enhancements include the **exponential transform**:

$$f(x) = a\exp(bx) + c,$$

which provides greater enhancement at high digital numbers than at low digital numbers, and the **logarithmic transform**:

$$f(x) = a\log(x) + b,$$

which has the converse effect.

See also *histogram matching*.

Convolution operator A linear spatial operation performed on an image, in which the new *digital number* $I'(i,j)$ assigned to the *pixel* with coordinates (i,j) is calculated as a weighted sum of the digital numbers I of the pixels in the neighbourhood of (i, j). A general convolution operator can be expressed as

$$I'(i,j) = \sum_{k=a}^{b} \sum_{l=c}^{d} w(k,l)I(i+k,j+l),$$

where the matrix $w(k,l)$ of weights is called the **kernel**.

The operation defined by the above equation is the convolution of the image with the function defined by the matrix $w(k,l)$. An alternative method of implementing a convolution filter is to calculate the *Fourier transform* of the image, to multiply this by the Fourier transform of the matrix $w(k,l)$, and then to perform the inverse Fourier transform. In general it is more efficient to use direct convolution if the matrix $w(k,l)$ has a small spatial extent. When $w(k,l)$ has a large spatial extent its Fourier transform occupies a smaller region of Fourier transform (spatial frequency) space, and it is more efficient to use the Fourier transform method.

See *edge detection, line detection, shape detection, sharpening, smoothing*.

Co-polarisation A *radar* system operates in co-polarised mode if it detects radiation having the same *polarisation* as it transmits, for example HH-polarised or VV-polarised (see *HH-polarisation, VV-polarisation*). Compare *cross-polarisation*.

Coriolis parameter See *ocean currents and fronts*.

Corner-cube reflector See *radar transponder*.

Corona Series of U.S. military reconnaissance satellites, operating between June 1959 and May 1972. The satellites carried panchromatic cameras (Keyhole) giving resolutions of 8 m (to December 1963), 3 m (August 1963 to October 1969) and 2 m (September 1967 to May 1972). The data are now declassified.

URL: http://edcwww.cr.usgs.gov/dclass/dclass.html

Covariance matrix The covariance matrix C_{ij} of an N-band image (or part of an image) is defined as

$$C_{ij} = \frac{1}{n} \sum_{p=1}^{n} (I_i(p) - \langle I_i \rangle)(I_j(p) - \langle I_j \rangle)$$

for $i = 1$ to N and for $j = 1$ to N, where $I_i(p)$ is the digital number of pixel number p in band i, $I_j(p)$ is the digital number of pixel number p in band j, and the image contains n pixels. $\langle I_i \rangle$ is defined as the mean digital number in band i:

$$\langle I_i \rangle = \frac{1}{n} \sum_{p=1}^{n} I_i(p)$$

and similarly for $\langle I_j \rangle$.

An element C_{ii} on the leading diagonal of the matrix is the variance σ_i^2 of the digital numbers in band i. An off-diagonal term C_{ij} ($j \neq i$) is related to the correlation coefficient ρ_{ij} between bands i and j through

$$C_{ij} = \rho_{ij}\sigma_i\sigma_j.$$

Critical angle See *Fresnel coefficients*.

Crop marks in archaeology The most sensitive method for detecting buried archaeological structures is based on the response of growing plants to differences in humidity. Plant height and colour are affected when soil moisture is limited. Detection via crop marks has been responsible for the discovery of more archaeological sites than all other methods combined. Being a consequence of the interaction of growing vegetation, soil structure and climatic change, it is hard to analyse. Growth is either retarded or advanced when a dry spell causes depletion of soil moisture reserves and plants must acquire their moisture from lower levels. Drainage is also a factor.

Markings over buried ditches and pits are more common than those over walls which have a negative effect on crop growth. Feature contrast is usually quite weak, and is visible only when photographed at optimum angles at low altitudes. When grain crops ripen, visible contrast rises if moisture stress conditions continue, and lines of deep green stand out sharply against a yellow background. After ripening, growth changes may be permanent and features may be seen as shadows in oblique illumination. Sites are visible for a week or more, unless differences become permanent at the end of the growing season prior to harvesting.

See also *archaeological site detection*.

Crossover A point at which the sub-satellite track of a satellite orbit intersects itself. Crossovers are particularly important in *radar altimeter* observations since they permit the orbital parameters to be calculated more accurately.

Cross-polarisation A *radar* system operates in cross-polarised mode if the polarisation states of the transmitted and received radiation are different, for example HV-polarised or VH-polarised (see *HV-polarisation*, *VH-polarisation*). Compare *co-polarisation*.

CRSS See *Carterra-1*.

Cryogenic Limb Array Etalon Spectrometer See *CLAES*.

CSA The Canadian Space Agency, officially established in 1990.

URL: http://www.space.gc.ca/welcomee.html

CSIRO (Commonwealth Scientific and Industrial Research Organisation) The Australian organization with responsibility for space research. See *CEOS*.

Cubic interpolation (cubic convolution) See *resampling*.

CZCS (Coastal Zone Color Scanner) U.S. optical/near infrared/thermal infrared imaging radiometer, carried on *Nimbus-7* satellite. Wavebands: 433–453, 510–530, 540–560, 660–680, 700–800 nm, 10.5–12.5 μm. Spatial resolution: 825 m at nadir. Swath width: 1600 km.

CZCS was primarily intended for *ocean colour* (chlorophyll and suspended sediment) measurements.

URL: http://eosdata.gsfc.nasa.gov/SENSOR_DOCS/CZCS_Sensor.html

DAPP See *DMSP*.

DARA Deutsche Agentur für Raumfahrtangelegenheiten, the German Space Agency.

Data Collection System (DCS) General name for a satellite-based system for the interrogation, storage and relay of data from automatic data-logging systems on the Earth's surface, such as the *Argos* system. DCS is now included on many remote-sensing satellites.

DCS See *Data Collection System*.

Debye equation The Debye equation describes the variation of the *dielectric constant* of a simple organic material containing polar molecules. It is normally written as follows:

$$\varepsilon' = \varepsilon_\infty + \frac{\varepsilon_p}{1 + \omega^2 \tau^2},$$

$$\varepsilon'' = \frac{\omega \tau \varepsilon_p}{1 + \omega^2 \tau^2},$$

where ε_∞ is the value of ε' at frequencies much greater than $1/\tau$, ε_p is the contribution to the dielectric constant from the polar molecules, ω is the angular frequency and τ (the relaxation time) is a measure of the time taken for the molecules to respond to a change in direction of the electric field.

The Debye equation provides a good model of the dielectric constant of water in the microwave region of the electromagnetic spectrum.

Decibel A logarithmic unit defining the ratio of two powers, intensities, radiances etc. A signal of power P_1 exceeds one of power P_2 by

$$10 \log_{10} \frac{P_1}{P_2} \text{ decibels.}$$

Defense Meteorological Satellite Program See *DMSP*.

DELTA-2 Russian scanning passive microwave radiometer, planned for inclusion on *Okean-O* satellites. Frequencies: 6.6, 13.6, 22.2, 37.7 GHz. Spatial resolution: 100 km at 6.6 GHz to 20 km at 37.7 GHz. Swath width: 800 km.

DEM See *digital elevation model.*

Dense medium model A volume *scattering model* that is valid for a dense discrete random medium, i.e. one in which scatterers are less than one wavelength apart. The scattering from nearby scatterers is thus correlated, and they act as a group rather than individually. The phase relation between scatterers and the average spacing between adjacent scatterers are important considerations for such media. The *radiative transfer model* originally developed for sparse media can be adapted to a dense medium by deriving a phase function for a unity volume of scatterers and allowing for near-field interactions among the scatterers. In a natural random medium the phase relation among scatterers is usually destroyed to a large extent by variations in size, shape and orientation of scatterers. However, the average spacing between the scatterers is not affected and hence must be taken into account.

Density slicing A very simple image *classification* procedure, applied to a single-band image or to one band of a multi-band image, in which ranges (slices) of *digital numbers* are assigned to particular digital numbers in the image display. Density slicing can be regarded as a *contrast enhancement* or as a one-dimensional *parallelepiped classifier*. If the 'slices' of a single-band image are displayed in different colours, the result is called a **pseudocolour image**.

Density slicing is commonly used where the image digital numbers have a direct relationship to a physical parameter of interest (for example, thermal infrared radiance may correspond directly to sea-surface temperature). It is also used to mask out regions of an image from further processing, and to reduce the effects of noise in an image.

Descending node The point in a satellite's *orbit* when it crosses the Earth's equatorial plane from the northern to the southern hemisphere.

Deserts and desertification Deserts are normally defined as regions of the Earth's land surface experiencing less than 3.6 metres of rainfall in a 30-year period and exhibiting evidence of degradation to plants and soil. Desertification is the anthropogenic change of (potentially) productive land to give desert-like conditions. Remote sensing-based methods for monitoring desertification normally make use of *vegetation mapping* techniques, but the use of *passive microwave* methods has also been shown to be effective. In particular, the horizontally and vertically polarised 37 GHz channels of the *SSM/I* instrument show a difference of approximately 30 K in brightness temperature over bare surfaces, whereas this difference is close to zero over vegetated surfaces. See also *erosion.*

Destriping Correction for the effects of *banding* in a poorly calibrated image acquired by a *scanning system*, especially *Landsat MSS*. Most destriping algorithms work by adjusting the *digital numbers* of a single strip of pixels so that the mean and standard deviation match the mean and standard deviation of a reference strip. For example, a typical destriping algorithm applied to Landsat MSS imagery, which has a banding period of six pixels, processes the image in blocks 100 pixels wide and 6 pixels in the along-track direction. The mean μ_i and standard deviation σ_i of the digital numbers in each strip ($i = 1$ to 6) of 100 pixels is calculated, and the digital numbers in strips 2 to 6 are transformed as follows:

$$I_i' = \frac{\sigma_1 I_i}{\sigma_i} + \mu_1 - \frac{\sigma_1 \mu_i}{\sigma_i},$$

where I_i is the original digital number, and I_i' the transformed value, of a pixel in strip i ($i = 2$ to 6).

Detector A device for converting electromagnetic radiation into an electrical signal. Detectors can be classified on the basis of the physical mechanisms that cause the conversion of radiation to signal. Photon detectors produce a signal when the mobility or number of free charge-carriers is changed by incident photons, and thermal detectors produce a signal when their temperature is changed by incident radiation. At optical and infrared wavelengths, the commonly used detectors are lead sulphide, indium antimonide, mercury or cadmium telluride, photoelectric detectors (see *photodiode*), thermopiles and thermistor bolometers. These all provide a near-linear relationship between radiance and electrical signal, but vary in their sensitivity to different parts of the spectrum, their ruggedness, and their response time.

Developing countries Some of the most successful, or potentially successful, applications of remote sensing are in developing countries. For some areas of some countries topographic maps are either non-existent or they are grossly lacking in detail or they contain serious errors. Satellite remote sensing therefore provides an important source of data that enables maps to be made or updated reasonably quickly, where conventional field survey or even aerial photography would be slow, tedious and expensive. There is, however, a problem in terms of major cost if the work is done outside the country. There may be a problem of training and technology transfer if the work is to be done within the country. A similar situation applies to geological maps as to topographic maps. Some developing countries are very sensitive to the fact that remotely sensed data and sophisticated technology may provide better information about their country's natural resources to a foreign company or foreign government than is available within the country itself.

In regard to applications of remote sensing beyond the field of mapping, such as resources monitoring, disaster monitoring, change detection, yield prediction etc., there are often problems in:

(a) making politicians and administrators aware of what are the potential uses of remote sensing
(b) access to remotely sensed data
(c) access to the data in near-real-time where appropriate (e.g. in monitoring floods or other disasters)

(d) availability of hardware and software systems for processing and interpreting the data

(e) availability of trained manpower with relevant expertise to analyse and interpret the data

(f) mechanisms for distribution of results to end users and, where necessary, in a timely fashion.

It is impossible to generalise about remote sensing in developing countries. Some developing countries have sophisticated technical installations for receiving and handling satellite data, some are building and launching their own satellites and some have extensive remote sensing applications programmes in hand, in some cases without sophisticated equipment or software but just relying on simple photo-interpretation techniques applied to hard-copy images. On the other hand, there are instances where people in a developing country have been sold expensive technology that is either inappropriate, or for which the infrastructure and skilled indigenous manpower is not available to enable it to be used properly. There are regional activities, such as those of the Asian Remote Sensing Society, and activities of some international organisations (FAO, the UN Outer Space Affairs Division, for example), which work very hard to ensure that developing countries are aware of the possibilities of remote sensing and have good opportunities to exploit the techniques.

See also *legal and international aspects*.

DFA (Dual-frequency altimeter) French dual-frequency radar altimeter, proposed for inclusion on *Jason* satellites. Frequency: C band (5.3 GHz) and K_u band (13.8 GHz). Pulse length (uncompressed): 105 µs; (compressed): 3.1 ns. PRF: 300 Hz (C band), 1800 Hz (K_u band). Beam-limited footprint: 450 km. Pulse-limited footprint: 2.2 km.

The DFA is derived from the *SSALT* instrument.

DFT See *Fourier transform*.

DIAL See *lidar*.

Dielectric constant The relative electric permittivity of a medium, usually denoted by the symbol ε_r. If the medium is loss-free, the dielectric constant is real; if it absorbs electromagnetic radiation, the dielectric constant is complex. Its real and imaginary parts are then usually written as

$$\varepsilon_r = \varepsilon' - i\varepsilon'',$$

where $i^2 = -1$. For a non-magnetic material (i.e. one for which the magnetic *permeability* $\mu_r = 1$), the dielectric constant is given by $\varepsilon_r = n^2$, where n is the *refractive index*. If

$$n = n' - i\kappa, \text{ then}$$

$$\varepsilon' = n'^2 - \kappa^2 \text{ and}$$

$$\varepsilon'' = 2n'/\kappa.$$

Differential absorption lidar See *lidar*.

Diffraction A change in the direction of propagation of a wave as a result of some obstruction of the wavefront, for example by the presence of an aperture, an opaque screen, or a phase-changing medium. Plane parallel electromagnetic radiation of wavelength λ incident on an aperture of width D is diffracted into a range of directions spanning an angular width of roughly λ/D radians, setting a limit on the spatial *resolution* of observing systems.

Diffuse albedo See *bidirectional reflectance distribution function*.

Diffuse illumination Electromagnetic radiation propagating over a wide range of directions, especially isotropically distributed radiation.

Diffusion equation (thermal) The conductive flow of heat inside a material is governed by the thermal diffusion equation:

$$\nabla^2 T = \frac{1}{\Gamma}\frac{\mathrm{d}T}{\mathrm{d}t},$$

where T is the temperature and Γ is the thermal diffusivity, defined as

$$\Gamma = \frac{K}{\rho c}.$$

K is the thermal conductivity of the material, ρ is its density, and c is its heat capacity per unit mass. Thermal diffusivity is the most convenient measure of the ability of a material to transfer heat from the surface sensed (e.g. a vegetation canopy, soil or lake surface) to the interior during heating, or conversely. Materials with low thermal diffusivity include water and sandy soil; limestone and gravel have intermediate thermal diffusivities; and quartzite and dolomite have high values of thermal diffusivity. See *thermal inertia*.

Digital elevation model (DEM, also called digital terrain model) A representation of the surface topography of a region of the Earth's surface, normally in *vector format* (i.e. stored like an image). DEMs are often obtained by digitising map contours, or by stereo matching aerial photographs or satellite images (see *stereophotography*). They can be used in the process of image *classification*, either as an extra 'feature' or, more commonly, to allow for the correction of atmospheric path and differential illumination effects.

Digital number (DN) The value associated with a *pixel* in a digital image, corresponding to the value of some physical quantity such as the *radiance* in a particular spectral band, measured at the detector. A pixel may have several digital numbers associated with it, for example in the case of a multispectral image. The digital number, also often called the grey level, is usually represented as a sequence of n binary digits, giving a range of possible values from 0 to $2^n - 1$.

Directivity A measure of the narrowness of the *power pattern* of an *antenna*. The directivity is defined as

$$D = \frac{4\pi}{\int P_n(\theta, \phi)\, d\Omega},$$

where $P_n(\theta, \phi)$ is the normalised power pattern, $d\Omega$ is an element of solid angle, and the integration is performed over 4π steradians. It is related to the *effective area* A_e of the antenna by

$$D = \frac{4\pi A_e}{\lambda^2},$$

where λ is the wavelength, and to the forward *gain G* by

$$G = \eta_1 D,$$

where η_1 is the antenna's radiation efficiency. If θ_0 and ϕ_0 are the *half-power beam widths* of the antenna in two orthogonal planes, both measured in degrees, the directivity is given approximately by

$$D \approx \frac{41\,253}{\theta_0 \phi_0}.$$

As with the forward gain, the directivity is often expressed in *decibels*.

Disaster monitoring The advent of remote sensing technology and spaceborne sensors in the 1960s made it possible to observe from space many of the Earth's vital signs and to monitor both human-induced and natural disasters. Disasters in progress as well as their aftermath are now routinely observed using instruments that sense in various parts of the electromagnetic spectrum. Droughts, floods, severe storms, pestilence, fires, volcanoes, nuclear power plant accidents, oil spills, and disappearing inland seas, have all been monitored and studied using remote sensing technology. The use of data from meteorological satellites for warnings of approaching cyclones and typhoons has probably already saved thousands of lives over the past three decades. Detecting these phenomena before they become disasters, and being able to predict their behaviour, is a goal to strive for in future decades.

See also *quality of life*.

Discrete Fourier transform See *Fourier transform*.

Discriminant function See *supervised classification*.

Dispersion

1. Dispersion of a wave is the phenomenon of waves of different frequencies travelling at different speeds in the same medium. A medium in which waves of different frequency travel at the same speed (for example, electromagnetic waves propagating in free space) is said to be non-dispersive. The dispersive properties of a wave are usually specified by the dispersion equation, which expresses the variation of the angular frequency ω as a function

of the wave number k, or equivalently, in the case of electromagnetic radiation, by the variation of the *refractive index* n as a function of the free-space wavelength λ. The phase velocity v_p is the speed at which crests and troughs of the wave are propagated. It is given by

$$v_p = \frac{\omega}{k} = \frac{c}{n},$$

where c is the speed of light *in vacuo*. The group velocity v_g is the speed at which any modulation applied to the wave (for example, pulses) is propagated. It is given by

$$v_g = \frac{d\omega}{dk} = \frac{c}{n - \lambda \dfrac{dn}{d\lambda}}.$$

2. Separation of electromagnetic radiation into components of different frequency. See *spectral resolution*.

Distortion See *geometric correction*.

Divergence See *separability*.

D layer See *ionosphere*.

DLR Deutsche Forschungsanstalt für Luft- und Raumfahrt, the German aerospace research establishment.

URL: http://www.dlr.de/

DMSP (Defense Meteorological Satellite Program) System of U.S. military satellites operated by the U.S. Air Force and formerly known as DAPP (Data Acquisition and Processing Program). About 30 DMSP satellites have been launched since 1966; the current generation is known as the 'Block 5D' satellites. Block 5D-1 satellites had the following launch and termination dates: F1 (September 1976 to September 1979), F2 (June 1977 to March 1978), F3 (April 1978 to December 1979), F4 (June 1979 to August 1980), F5 failed on launch in 1980. Block 5D-2 satellites had the following launch and termination dates: F6 (December 1982 to August 1987), F7 (December 1983 to October 1987), F8 (June 1987 to August 1991), F9 (February 1988), F10 (December 1990), F11 (November 1991), F12 (August 1994), F13 (March 1995). A further block, 5D-3, is planned. Objectives: Meteorological observations (global cloud cover, atmospheric temperature and humidity, ozone). Orbit: Circular *Sun-synchronous LEO* at 854 km altitude. Period 102.1 minutes; inclination 98.8°. Principal instruments: *OLS*, *SSC* (F4 only), *SSD* (F4 only), *SSH*, *SSH-2* (F6 only), *SSM/I* (F8 onwards), *SSM/T* (F4, F7 onwards). The programme provides for two satellites in orbit at any one time, with equatorial crossing times 12 hours apart. The satellites also carry solar–terrestrial physics instruments.

URL: http://web.ngdc.noaa.gov/dmsp/dmsp.html

DN See *digital number*.

Doppler effect The change in frequency of a wave as a result of relative motion of the transmitter and receiver (and, in the case of a sound wave, of the transmitting medium). For electromagnetic radiation *in vacuo*, the Doppler effect can be expressed as

$$\frac{\nu_r}{\nu_t} = 1 - \frac{(\mathbf{v}_t - \mathbf{v}_r)\cdot(\mathbf{x}_t - \mathbf{x}_r)}{c|\mathbf{x}_t - \mathbf{x}_r|},$$

where ν_t is the transmitted frequency, ν_r is the received frequency, $\mathbf{v}_t$ and $\mathbf{v}_r$ are the vector velocities of the transmitter and receiver, and $\mathbf{x}_t$ and $\mathbf{x}_r$ are the vector positions of the transmitter and receiver. c is the speed of light.

Doppler Orbitography and Radiopositioning Integrated by Satellite See *DORIS*.

DORIS (Doppler Orbitography and Radiopositioning Integrated by Satellite) Precise orbit-location system carried on *SPOT-2* to *-4* and *Topex-Poseidon* and planned for inclusion on *Envisat* and *Jason-1*. Signals are transmitted from ground stations at 0.4 and 2.0 GHz, and the Doppler shifts of the signals detected at the satellite are used to infer its position to an accuracy of 10–20 cm.

DTM See *digital elevation model*.

Dual Frequency Altimeter See *DFA*.

Dynamical form factor A dimensionless quantity representing the effect of the Earth's oblateness on its gravitational potential. The dynamical form factor is also called the second zonal harmonic, and usually given the symbol J_2. It is defined by the expansion of the Earth's gravitational potential V as a function of latitude ϕ and distance r from the Earth's centre:

$$V = -\frac{GM}{r}\left(1 - \frac{J_2 a_e^2}{2r^2}[3\sin^2\phi - 1] + \cdots\right),$$

where G is the universal gravitational constant, M is the Earth's mass, and a_e is the Earth's equatorial radius.

EarlyBird U.S. satellite launched in December 1997, operated by *EarthWatch* Inc., with a design life of 5 years. <u>Objectives</u>: High resolution Earth observation for commercial use. <u>Orbit</u>: Circular *Sun-synchronous LEO* at 470 km altitude. <u>Inclination</u> 97.3°. Equator-crossing time 10:30 descending. Communication with EarlyBird was lost almost immediately after launch, and the satellite is no longer functional.

EarlyBird carried a high-resolution optical/near infrared imager, operating in panchromatic and multispectral modes. <u>Wavebands</u>: 0.42–0.70 μm (panchromatic); 0.49–0.60, 0.62–0.67, 0.79–0.88 μm (multispectral). <u>Spatial resolution</u>: 3 m (panchromatic); 15 m (multispectral). <u>Image size</u>: 3 km × 3 km (panchromatic), 15 km × 15 km (multispectral), plus strip images up to 900 km².

The EarlyBird sensor's field of view could be steered by ±30° fore and aft, ±28° side to side, to provide stereo imaging and greater flexibility in coverage.

Earth Imaging System See *EIS*.

Earth Observing Scanning Polarimeter See *EOSP*.

Earth Observing System See *EOS*.

Earth radiation budget The balance between the solar radiation intercepted by the Earth, and the radiation reflected and re-emitted by the Earth. Incoming solar radiation is predominantly in the ultraviolet, optical and near infrared regions of the electromagnetic spectrum (**short-wave radiation**). On average, approximately 30% of this radiation is reflected in the same range of wavelengths (see *albedo*), mostly by *clouds*. The remaining 70% is absorbed by the Earth's surface (land and sea), the atmosphere, and clouds, and re-emitted at longer (predominantly thermal infrared) wavelengths (**long-wave radiation**). The figure summarises, in a simplified form, the Earth's spatially and temporally averaged radiation budget. In fact, the radiation budget exhibits large spatial and temporal variations.

The lighter arrows represent short-wave radiation and the darker arrows long-wave radiation. The units are percentages of the spatially and temporally averaged solar radiation intercepted by the Earth. The four components of the long-wave radiation emitted by the Earth's surface are evaporation and precipitation (23%), atmospheric absorption (15%), sensible heat transfer (7%) and emission to space (6%).

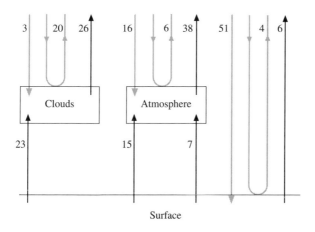

Understanding of the Earth radiation budget is a primary requirement for understanding and modelling of the global *climate* system, and has received increased prominence as a result of concerns about global climate change. It is currently possible to measure both the short-wave and long-wave budgets at the top of the atmosphere, the short-wave budget at the Earth's surface, and the total incoming radiation flux, integrated typically over a range of wavelengths from 0.2 to 4.0 μm.

Measurements of the Earth radiation budget from spaceborne sensors were made using data from the early *TIROS*, *ITOS* and *Nimbus* satellites, giving the first global estimates of short-wave and long-wave fluxes. The Earth Radiation Budget Experiment (*ERBE*) was a dedicated programme of three satellites launched in the mid-1980s, which provided the most comprehensive data set of albedo and long-wave fluxes. The principal limitations of this data set are its limited spatial and temporal sampling frequencies and its absolute accuracy. The next generation of Earth radiation budget instruments, notably *CERES*, should address these problems.

URL: http://nssdc.gsfc.nasa.gov/earth/rb.html

Earth Radiation Budget Experiment See *ERBE*.

Earth Radiation Budget Satellite See *ERBS*.

Earth Radiation Budget Sensor See *ERB*.

Earth Resources Experimental Package See *EREP*.

Earth Resources Technology Satellite See *Landsat 1–3*.

Earthwatch Imager U.S. optical/near infrared imaging *CCD* (*step-stare*) radiometer, planned for inclusion on *Clark*. Wavebands: Panchromatic: 0.45–0.80 μm; Multispectral: 0.50–0.59, 0.61–0.68, 0.79–0.89 μm. Spatial resolution: 3 m (panchromatic); 15 m (multispectral). Swath width: 12 km.

EarthWatch Inc Commercial satellite imaging company. Address: 1900 Pike Rd, Longmont, CO 80501-6700, U.S.A..

URL: http://www.digitalglobe.com

Eccentricity The extent to which a satellite *orbit* departs from circularity, defined by the equation

$$e = \sqrt{1 - b^2/a^2},$$

where e is the eccentricity, and a and b are respectively the semimajor and semiminor axes of the elliptical orbit.

Edge detection The process of identifying the pixels in an image that form the boundaries between homogeneous regions, homogeneity normally being defined radiometrically. Edge detection is often a precursor to image *segmentation*.

Edge pixels are usually detected by applying *convolution operators* that calculate components of the spatial gradient of the image brightness. If the output from such an operator exceeds a suitable threshold, it is assumed that the pixel forms part of an edge having a particular orientation. If the orientation is of no concern, the pixel is identified as an edge pixel if the sum of the squares of the outputs from two operators corresponding to orthogonal orientations exceeds a suitable threshold.

Examples of common edge-detection kernels are given below:

Roberts operators

```
 0  1      -1  0
-1  0       0  1
```

Sobel operators

```
-1  0  1      1  2  1
-2  0  2      0  0  0
-1  0  1     -1 -2 -1
```

The **Laplacian** operator finds the isotropic second derivative, and is therefore insensitive to the orientation of the edge. It has the following kernel:

```
0  1  0
1 -4  1
0  1  0
```

Since edge-detection filters remove the low spatial frequencies from an image, they are sometimes referred to as **high-pass filters**.

Edge enhancement See *sharpening*.

Effective area The effective area A_e of an *antenna* is defined such that the power received per unit frequency interval is

$$\frac{F_\nu A_e}{2}$$

when electromagnetic radiation of spectral *flux density* F_ν is incident upon it. See *power pattern, directivity*.

EGS (Experimental Geodetic Satellite) Japanese satellite carrying *laser retroreflectors* which can be ranged from the ground for geodetic purposes. Launched August 1986. Orbit: semimajor axis 7900 km, inclination 50°, period 116 minutes.

Einstein The Einstein unit is a mole (6.02×10^{23}) of photons. See *photosynthetically active radiation*.

EIS (Earth Imaging System) System of two *CCD* imaging cameras (WAC = wide-angle camera; NAC = narrow-angle camera) carried on *PoSAT-1*. Wavebands: 0.81–0.89 μm (WAC); 0.61–0.69 μm (NAC). Spatial resolution: 2 km (WAC); 200 m (NAC). Image size: 1500 km × 1000 km (WAC); 150 km × 100 km (NAC).

E layer See *ionosphere*.

Electrically Scanning Microwave Radiometer See *ESMR*.

Electromagnetic bias A systematic error observed in *radar altimeter* measurements over ocean surface, in which the mean ocean height is underestimated by an amount depending on the *significant wave height*. It arises through a height asymmetry in the distribution of radar backscatter coefficient about the mean surface level, and has a magnitude of approximately 2–3% of the significant wave height. Electromagnetic bias is also known as **sea-state bias**.

Electromagnetic radiation System of oscillating electric and magnetic fields that carry energy from one point to another. The electric and magnetic fields are oriented at right angles to each other and to the direction of propagation. Electromagnetic radiation can propagate in a vacuum, in which case the speed of propagation is equal to the speed of light *in vacuo*, or in suitable media. See *Maxwell's equations*.

Electromagnetic spectrum The range of frequencies (or equivalently wavelengths) over which electromagnetic radiation can be propagated. For convenience, it is divided into a number of frequency bands. The following table shows the internationally accepted (except *, which are definitions adopted in this book) names for those parts of the spectrum commonly used in remote sensing:

Band	Frequency range	Wavelength range
HF (High Frequency)	3–30 MHz	10–100 m
VHF (Very High Frequency)	30–300 MHz	1–10 m
UHF (Ultra-High Frequency)	300 MHz–3 GHz	0.1–1 m
P band (microwave)	225–390 MHz	0.77–1.33 m
L band (microwave)	390 MHz–1.55 GHz	0.19–0.77 m
SHF (Super-High Frequency)	3–30 GHz	10–100 mm
S band (microwave)	1.55–5.2 GHz	58–190 mm
X band (microwave)	5.2–10.9 GHz	28–58 mm
K band (microwave)	10.9–30.6 GHz	8.3–28 mm
K_u band (microwave)	15.4–17.3 GHz	17–20 mm
K_l band (microwave)	15.4–24.5 GHz	12–17 mm
EHF (Extremely High Frequency)	30–300 GHz	1–10 mm
K_a band (microwave)	33–36 GHz	8.3–9.1 mm
Q band (microwave)	36–46 GHz	6.5–8.3 mm
V band (microwave)	46–56 GHz	5.4–6.5 mm
W band (microwave)	56–100 GHz	3.0–5.4 mm
IR (Infrared)	300 GHz–385 THz	0.78 μm–1 mm
Far infrared*	300 GHz–10 THz	30 μm–1 mm
Thermal infrared*	10–100 THz	3.0–30 μm
Near infrared*	100–385 THz	0.78–3.0 μm
VIS (Visible light)	385–789 THz	0.38–0.78 μm
UV (Ultraviolet)	789 THz–30 PHz	0.01–0.38 μm

See also *long-wave radiation, short-wave radiation*.

Electro-optical sensors Electro-optical sensors record electromagnetic radiation using a *detector* operating in the optical to thermal infrared region. They are currently the most commonly used sensors in remote sensing. The two types of electro-optical sensor are non-imaging (e.g. *radiometers*) and imaging (e.g. scanner, *CCD*, framing camera, *vidicon*). Non-imaging sensors do not produce a two-dimensional representation of the upwelling radiance (i.e. a 'picture'), but rather integrate radiance measurements over portions of time, space, waveband and solar and sensor geometries for the area within the sensor's field of view. Imaging sensors produce a 'picture' of the Earth, its oceans and atmosphere over portions of time, space, waveband and solar and sensor geometries but do so for each ground resolution element.
See also *scanning*.

Elektro See *GOMS*.

Ellipsoid The simplest model of the shape of the Earth that takes into account its departure from sphericity is an ellipsoid of revolution with rotational symmetry about the polar axis. The ellipsoid is characterised by the length $2a$ of its longer (equatorial) axis and the length $2b$ of its shorter (polar) axis. Historically, geodetic measurements to determine a and b were made over limited arcs of the Earth's surface, and these 'locally correct' ellipsoids often still form the basis of national *map projections*. However, more recent satellite-based determinations provide globally fitted values of a and b which often form the basis of projections used for satellite imagery.

The table below lists the parameters a and b (in metres) of some of the main ellipsoids.

Ellipsoid	Date	a	b	Main region
Airy	1830	6 377 563	6 356 257	Britain, Ireland
Everest modified	1830	6 377 301	6 356 100	India
Bessel	1841	6 377 397	6 356 079	Central Europe
Bessel modified		6 377 492	6 356 174	
Airy modified	1849	6 377 340	6 356 034	
Clarke	1866	6 378 206	6 356 584	North America
Clarke	1880	6 378 249	6 356 515	France, Africa
International	1924	6 378 388	6 356 912	Global
Krasovsky	1940	6 378 245	6 356 863	Russia, Eastern Europe
IAU	1967	6 378 160	6 356 775	Global, Australia
WGS	1972	6 378 135	6 356 751	Global
WGS*	1984	6 378 137	6 356 752	Global

*Also GEM-10C, GRS 1980

See also *geoid*.

Elliptical polarisation See *polarisation*.

Ellipticity See *polarisation*.

Emissivity The ratio of the thermally generated power, flux density, radiance etc. emitted by a body at temperature T to the power etc. that would be emitted by that body at the same temperature T if it were a *black body*. Emissivity is usually denoted by the symbol ε or e, and is often specified as a function of wavelength or frequency.

The emissivity of a perfectly smooth surface is given by

$$\varepsilon = 1 - |r|^2,$$

where r is the *Fresnel coefficient* for radiation reflected from the surface at the same wavelength, in the same direction and with the same polarisation as the emitted radiation. For a perfectly rough (Lambertian) surface, the emissivity is

$$\varepsilon = 1 - \frac{\gamma_0}{4}$$

independent of direction, where γ_0 is the value of the *backscatter coefficient* σ^0 at normal incidence. This is determined only by the dielectric constant of the surface.

In general, the emissivity of a surface can be written in terms of the *bidirectional reflectance distribution function* (BRDF) R as

$$\varepsilon_p = 1 - \int (R_{pp} + R_{qp}) \cos \theta_1 \, d\Omega_1,$$

where ε_p is the emissivity for polarisation state p, R_{qp} is the BRDF for radiation incident in polarisation state q and reflected in state p, R_{pp} is the BRDF for radiation incident and reflected in state p, θ_1 is the angle between the reflected radiation and the surface normal, and $d\Omega_1 = \sin \theta_1 \, d\theta_1 \, d\phi_1$ where ϕ_1 is the azimuth angle of the scattered radiation. The integral is taken over 2π steradians, i.e. $\theta_1 = 0$ to $\pi/2$ and $\phi_1 = 0$ to 2π. This expression can also be written in terms of the bistatic backscatter coefficient σ^0 as

$$\varepsilon_p = 1 - \frac{\int (\sigma_{pp}^0 + \sigma_{qp}^0) \, d\Omega_1}{4\pi \cos \theta_0},$$

where θ_0 is the angle between the incident radiation and the surface normal.

Enhanced Thematic Mapper See *ETM*.

Environmental Satellite See *Envisat*.

Environmental Science Service Administration See *ESSA*.

Envisat (Environmental Satellite) European Satellite, operated by *ESA*, scheduled for launch in 1999 with a nominal lifetime of 5 years. Originally known as Columbus Polar Platform. Envisat will be part of the *POEM* mission. Objectives: Observation and measurement of: vegetation and soil moisture; snow and ice extent; fires; marine chlorophyll, sediment, and pollution; ocean wind speeds, wave spectra and SWH; sea ice; coastal erosion; clouds; atmospheric chemical composition, aerosols and water vapour; Earth radiation budget. Orbit: Circular *Sun-synchronous LEO* at 785 km altitude. Period 100.6 minutes; inclination 98.54°; equator crossing time 10:00 (ascending node). Exactly-repeating orbit (501 orbits in 35 days). Principal instruments: *AATSR, ASAR, DORIS, GOMOS, MERIS, MIPAS, MWR, Radar Altimeter-2, Scarab, Sciamachy.*

URL: http://envisat.estec.esa.nl/

EOS (Earth Observing System) Planned system of satellites in *NASA*'s *Mission to Planet Earth* programme, with contributions from other U.S. and non-U.S. operators. The first satellite is scheduled for launch in 1998. The programme has a nominal lifetime of 25 years or more. Objectives: The principal emphasis of the programme will be on global change. Satellites: *ADEOS-II, EOS-AM, EOS-Chem, EOS-Laser Alt, EOS-PM, ISSA, Jason-1, Landsat-7, Meteor-3M-1, TRMM.*

URL: http://eospso.gsfc.nasa.gov/

58

Eos-Aero See *ISSA*, *Meteor-3M*.

EOS-Alt See *EOS-Laser Alt*, *Jason-1*.

EOS-AM (EOS Morning [= ante meridian] platform) Part of the *EOS* programme. EOS-AM satellites are scheduled for launch in 1998, 2004 and 2010, with nominal lifetimes of 6 years. Objectives: Measurements of clouds, aerosols, atmospheric trace gases, Earth surface, Earth radiation budget. Orbit: Circular *Sun-synchronous* LEO at 705 km altitude. Period 100 minutes; inclination 98.2°; equator crossing time 10:30 (descending node). Principal instruments: *ASTER* (AM-1), *CERES*, *EOSP* (AM-2 and AM-3), *LATI* (AM-2, AM-3), *MISR*, *MODIS*, *MOPITT* (AM-1).

EOSAT (Earth Observation Satellite Company) Company originally set up in 1984 to distribute *Landsat* data on a commercial basis. The company now distributes a wide range of remote sensing data. Address: 4300 Forbes Boulevard, Lanham, MD 20706-9954 U.S.A.

URL: http://www.eosat.com/

EOS-Chem (EOS Chemistry mission) Part of the *EOS* programme. EOS-Chem satellites are scheduled for launch in 2002 and 2008, with nominal lifetimes of 6 years. Objectives: Profiling of atmospheric (troposphere and lower stratosphere) ozone, water vapour and trace gases. Orbit: Circular *Sun-synchronous* LEO at 705 km altitude. Period 100 minutes; inclination 98.2°. Principal instruments: *HiRDLS*, *MLS*, *ODUS* (EOS-Chem-1), *TES*.

EOS-Laser Alt (EOS Laser Altimeter satellite) Part of the *EOS* programme. EOS-Laser Alt satellites are scheduled for launch in 2002 and 2008, with nominal lifetimes of 3 years. Objectives: Laser profiling of ice sheets, land surface and cloud tops. Orbit: Circular at 705 km altitude. Period 100 minutes; inclination 94°. Principal instruments: *GLAS*. The satellites will also carry *GPS* receivers for orbit determination.

EOSP (Earth Observing Scanning Polarimeter) Instrument planned for inclusion on the *EOS-AM* satellites of the *EOS* satellite system. Wavebands: 12 channels between 0.41 μm and 2.25 μm (orthogonal linear polarisations measured in each band). Spatial resolution: 10 km at nadir. Swath width: limb to limb.
EOSP is intended to provide characterisation of aerosol distributions (optical thickness between 0 and 35 km altitude), clouds (cloud-top pressure and scattering), and atmospheric correction radiances.

EOS-PM (EOS Afternoon [= post meridian] platform) Part of the *EOS* programme. EOS-PM satellites are scheduled for launch in 2000, 2006 and 2012, with nominal lifetimes of 6 years. Objectives: Measurements of clouds, precipitation, radiation flux, SST, ocean productivity, snow and sea ice. Orbit: Circular *Sun-synchronous*

LEO at 705 km altitude. Period 100 minutes; inclination 98.2°; equator crossing time 13:30 (ascending node). Principal instruments: *AIRS, AMSU/MHS, CERES, MIMR, MODIS.*

Equation of time See *solar illumination direction.*

Equilibrium line See *glaciers.*

Equivalent reflectance See *atmospheric correction.*

ERB (Earth Radiation Budget Sensor) U.S. ultraviolet/optical/infrared radiometer for Earth radiation budget measurements, carried on *Nimbus-6* and *-7*. Wavebands: 0.2–50, 0.2–3.6, 0.7–2.8 μm. Spatial resolution: 150 km to 1500 km. ERB also had 10 Sun-viewing bands between 0.2 and 50 μm.

ERBE (Earth Radiation Budget Experiment)
1. U.S. ultraviolet/optical/infrared radiometer for Earth radiation budget measurements, carried on *NOAA-9* and *-10*. Wavebands: 0.5–0.7, 0.2–4.0, 0.2–50, 10.5–12.5 μm. Spatial resolution: 50 km at nadir (scanning); 200 km at nadir (non-scanning). Swath width: 3000 km.
2. U.S. ultraviolet/optical/infrared radiometer for Earth radiation budget measurements, carried on *ERBS*. Wavebands: 0.2–50 μm (4 bands) nadir viewing, 0.2–5.0 μm (scanning), 5–50 μm (scanning), 0.2–50 μm (scanning). ERBE also had a Sun-viewing radiometer.

ERBS (Earth Radiation Budget Satellite) U.S. satellite, operated by *NASA*, launched in October 1984. Objectives: Measurement of the Earth's radiation budget. Orbit: Circular *LEO* at 610 km altitude. Period 97 minutes; inclination 57°. Principal instruments: *ERBE, SAGE II.* ERBS now operates only when in regions of sufficient sunlight.

URL: http://asd-www.larc.nasa.gov/erbe/erbs.html

EREP (Earth Resources Experimental Package) Package of six sensors carried on *Skylab.* The sensors were *S190A, S190B, S191, S192, S193* and *S194.*

Erosion, soil Soil erosion by wind and water is a major aspect of *desertification,* resulting in the loss of water-storage capacity, nutrients, and the soil itself. Methods for remote sensing of soil erosion are generally based on changes in the colour, *texture* and structure (for example the appearance or growth of characteristic drainage patterns) in high-resolution images such as *Landsat* images.

Error matrix A method, also called the confusion matrix, of characterising the performance of a *classification* technique. It is a square matrix of $n \times n$ elements, where n is the number of classes. The element a_{ij} is the number of pixels known to belong to class i and identified as belonging to class j. The **user's** (or **consumer's**) **accuracy** for class i is given by

$$\frac{a_{ii}}{\sum_{j=1}^{n} a_{ji}},$$

and the **producer's accuracy** is

$$\frac{a_{ii}}{\sum_{j=1}^{n} a_{ij}}.$$

Single-parameter classification accuracies can also be specified from the error matrix. The simplest of these,

$$\frac{\sum_{i=1}^{n} a_{ii}}{\sum_{i=1}^{n} \sum_{j=1}^{n} a_{ij}},$$

is just the proportion of pixels that are correctly classified. The κ-value (kappa value) is defined as

$$\kappa = \frac{\left(\sum_{i=1}^{n} \sum_{j=1}^{n} a_{ij}\right) \sum_{i=1}^{n} a_{ii} - \sum_{i=1}^{n} \left(\sum_{j=1}^{n} a_{ij} \sum_{j=1}^{n} a_{ji}\right)}{\left(\sum_{i=1}^{n} \sum_{j=1}^{n} a_{ij}\right)^2 - \sum_{i=1}^{n} \left(\sum_{j=1}^{n} a_{ij} \sum_{j=1}^{n} a_{ji}\right)}.$$

ERS-1 (European Remote Sensing Satellite) European satellite operated by *ESA*, launched in July 1991 with a nominal lifetime of 3 years or more. Objectives: ERS-1 was an experimental, pre-operational satellite for investigation of the Earth's land, ocean and ice surfaces, atmospheric physics and meteorology. Orbit: Circular *Sun-synchronous LEO* at 775 [781] km altitude. Period 100.5 [100.6] minutes; inclination 98.5°. Exactly repeating orbit (43 [501] orbits in 3 [35] days). Principal instruments: *AMI*, *ATSR*, *Radar Altimeter*. ERS-1 also carries a laser retroreflector for precise determination of its position (by laser ranging from ground stations), and the *PRARE* ranging system (which failed to become operational after launch).

The 3-day repeat orbit is the 'ice orbit', providing rapid revisits. The 35-day repeat is the 'mapping orbit' giving full coverage within the latitudinal limits of the orbit.

URL: http://www.esoc.esa.de/external/mso/ers.html

ERS-2 (European Remote Sensing Satellite) European satellite operated by *ESA*, launched in April 1995 with a nominal lifetime of 3 years. Objectives: Follow-on from *ERS-1*, with the addition of atmospheric chemistry observations. Orbit: Circular *Sun-synchronous LEO* at 775 km altitude. Period 100.5

minutes; inclination 98.5°; equator crossing time 09:30. *Exactly-repeating orbit* (43 orbits in 3 days). Principal instruments: *ATSR-2, AMI, GOME, Radar Altimeter*. ERS-2 also carries a laser retroreflector for precise determination of its position (by laser ranging from ground stations), and the *PRARE* ranging system.

URL: http://www.esoc.esa.de/external/mso/ers.html

ERTS See *Landsat 1–3*.

ESA The European Space Agency. An international organisation with membership from Austria, Belgium, Denmark, Finland, France, Germany, Ireland, Italy, the Netherlands, Norway, Spain, Switzerland, Sweden and the United Kingdom.

URL: http://www.esrin.esa.it/htdocs/esa/esa.html

ESMR (Electrically Scanning Microwave Radiometer) U.S. single-frequency electrically scanned passive microwave radiometer, carried on *Nimbus-5* and *-6*.
   Nimbus-5: Frequency: 19.35 GHz (bandwidth 250 MHz). Polarisation: H. Spatial resolution: 25 km at nadir. Swath width: 3000 km. Absolute accuracy: 2 K.
   Nimbus-6: Frequency: 37 GHz (bandwidth 250 MHz). Polarisation: H and V. Spatial resolution: 20 km at nadir. Swath width: 1300 km. Absolute accuracy: 2 K.

ESSA (Environmental Science Services Administration) Series of nine U.S. satellites operated by *NOAA* from February 1966 to November 1972. Objectives: Operational meteorological observations. Each satellite had a lifetime of typically 2 years. Orbit: Nominally circular *LEO*. With the exception of ESSA-1, the satellites had orbital heights between 1520 and 1650 km (minimum altitude) and 1640 and 1730 km (maximum altitude). ESSA-1 had a minimum altitude of 800 km and a maximum of 965 km. Period 100 minutes (ESSA-1), 113 to 115 minutes (ESSA-2 to -9); inclination 98° (ESSA-1), 101 to 102° (ESSA-2 to -9). Principal instruments: *APT* (even-numbered satellites), *AVCS* (odd-numbered), *FPR* (even-numbered).

Etalon Soviet satellites carrying *laser retroreflectors* which can be ranged from the ground for geodetic purposes. Launched January 1989 (Etalon-1), May 1989 (Etalon-2). Orbit: altitude 19 130 km, inclination 64.8°, period 675 minutes.

ETM (Enhanced Thematic Mapper) 8-band optical/infrared imaging radiometer to be included on board *Landsat-7* satellite. Wavebands: 0.50–0.90 μm (panchromatic band), 0.45–0.52 μm, 0.52–0.60 μm, 0.63–0.69 μm, 0.76–0.90 μm, 1.55–1.75 μm, 2.08–2.35 μm, 10.4–12.5 μm. Spatial properties from 700 km altitude: Spatial resolution: 15 m (panchromatic band), 60 m (thermal infrared band), 30 m (all other bands); swath width: 185 km. The ETM differs from the *TM* by virtue of the inclusion of the panchromatic band.

Strictly, the ETM was the instrument carried by Landsat-6, which failed to become operational. The Landsat-7 instrument is designated ETM+. NASA has proposed a new instrument **LATI** (Landsat Advanced Technology Instrument), which may be included on the *EOS-AM* satellites.

Euclidean distance See *clustering, supervised classification*.

EUMETSAT (European Organisation for the Exploitation of Meteorological Satellites) See *CEOS*.

URL: http://www.eumetsat.de

European Remote Sensing Satellite See *ERS-1, ERS-2*.

European Space Agency See *ESA*.

Exactly repeating orbit A satellite *orbit* in which the *sub-satellite point* traces a closed path on the Earth's surface. The condition for an exactly repeating orbit is

$$P_N(\Omega - \Omega_E) = 2\pi \frac{D}{N},$$

where P_N is the satellite's *nodal period*, Ω is the angular velocity of *precession* of its orbit, Ω_E is the angular velocity of the Earth's rotation about its axis ($=2\pi$ radians per *sidereal day*). D is an integer representing the number of days between repeats, and N is another integer representing the number of orbits (e.g. passages through the *ascending node*) between repeats. The ratio D/N is expressed in its lowest terms, so that D and N have no common factors. The condition for an exactly repeating *Sun-synchronous* orbit is more simply expressed as

$$P_N = 86\,400 \frac{D}{N},$$

where P_N is measured in seconds. Exactly repeating orbits are widely used for remote-sensing satellites, since they permit the locations from which data can be obtained to be specified in a simple way.

The value of N determines the spatial density of sub-satellite tracks on the Earth's surface, since spatially adjacent northbound sub-satellite tracks are separated by $360/N$ degrees of longitude (the same formula applies to spatially adjacent southbound tracks). Thus, a dense coverage (which would be required, for example, to achieve global coverage with a narrow-swath sensor) requires a large value of N. Since D/N is proportional to the nodal period P_N, which is in turn determined by the satellite's *semimajor axis*, a large value of N is likely to require a large value of D, giving a (possibly undesirably) long interval between successive views of the same point on the Earth's surface. Some flexibility can be obtained if the direction of the sensor's field of view can be varied.

The values of D and N determine the sequence in which spatially adjacent sub-satellite tracks are traced out. If the ratio D/N is sufficiently close to an integer ratio D'/N', where $D' < D$, $N' < N$ and D' and N' have no common factors, spatially adjacent sub-satellite tracks will be traced at intervals of D' days. This is referred to as a D'-day orbital **subcycle**, and the condition that must be satisfied can be written as

$$\frac{D'N - 1}{D} = \text{integer}.$$

If the sensor's swath width is sufficiently wide and D' is sufficiently small, such an orbit allows a given point on the Earth's surface to be viewed several times in quick succession, followed by a longer period during which it cannot be viewed.

Exitance See *radiant exitance*.

Experimental Geodetic Satellite See *EGS*.

Exponential contrast enhancement See *contrast enhancement*.

Extinction Synonymous with *attenuation*.

Facet model A surface *scattering model* that assumes that a rough surface may be modelled as a collection of facets, characterised by their size and slope distributions. Such a model is intuitively appealing, but quite difficult to apply in practice because of the difficulty of defining the facet size distribution. The presence of facet edges in the model (but not in reality) introduces error. If all the facets are large compared with the wavelength of the incident radiation, then they may be approximated by infinite planes. In this case the model reduces to the high-frequency limit of the *Kirchhoff model*.

Far infrared See *infrared*.

Fast Fourier transform See *Fourier transform*.

Feng Yun See *FY-1, FY-2*.

Fetch See *significant wave height*.

FFT See *Fourier transform*.

Field of view (FOV) The total area viewed by a *scanning* remote sensing system, as the instantaneous field of view is scanned over the surface. See *spatial resolution, swath width*.

Fire Detection and monitoring of fire (especially forest fires) can be performed using aerial photography or visible-wavelength scanner imagery during daylight conditions, and using thermal infrared imagery at night. The visible-wavelength techniques respond to the radiation emitted by the flames, or to the presence of smoke. Thermal infrared observations respond to the high temperature of the burning material. Thermal infrared imagery can also reveal the presence of thick plumes of smoke at some distance from a fire, as a result of the thermal contrast between the smoke-laden air and the background.

Flat-Plate Radiometer See *FPR*.

F layer See *ionosphere*.

Fluorescence Re-emission of absorbed radiation at a different (usually lower) frequency, without first converting the absorbed energy into heat, by electronic transitions in the atoms or molecules of the fluorescent material. Many minerals exhibit visible-wavelength fluorescence when illuminated by ultraviolet radiation. Chlorophyll in green plants exhibits fluorescence in the near infrared (0.7–0.85 µm) when illuminated by visible-wavelength radiation, although this is not the dominant contribution to the high near infrared reflectance of plant material.

Flux See *radiant flux*.

Fog See *clouds*.

Footprint Equivalent to the instantaneous field of view (IFOV) of a sensor (see *spatial resolution*). The term 'footprint' is generally used in preference to IFOV in the case of non-imaging sensors.

Forests Forests represent important natural resources and also play a significant role in the global climate system, through their contribution to the global carbon cycle and their effect on the global *albedo*. Remote sensing methods are well established for monitoring, classification and evaluation of forest areas. At the highest spatial resolutions (scales from 1:100 000 to 1:10 000), aerial photography can be used to identify individual trees, their height (from *relief displacement* or *stereophotography*) and crown area, and their state of health. Black and white photographs, both optical and near infrared, are still extensively used for this purpose, although colour photography improves the classification accuracy. Timber volume can be estimated from height and crown area using empirically determined and species-dependent relationships. At lower spatial resolutions, including those available from spaceborne *imaging radiometers*, individual trees can not be resolved, and analysis normally involves *classification* of the image into homogeneous areas combined with field identification and measurement of species, timber volume etc. The use of calibrated optical/near infrared imagery also permits the estimation of forest *biomass* through the use of *vegetation indices*. Forest damage by pests, diseases, air pollution, fire and storms can be identified by defoliation (especially using vegetation indices), colour change, and textural changes. Forest *fires* can be identified in visible-wavelength and thermal infrared images.

Increasing use is being made of *synthetic aperture radar* imagery for the characterisation and monitoring of forest vegetation.

Fourier transform If some quantity f, which may be real, imaginary or complex, varies with time, it can be expressed as a function of time as $f(t)$. It can equivalently be expressed as the sum of sinusoidal and cosinusoidal components having different amplitudes and angular *frequencies* ω. In general, the sum must be carried out over an infinite number of terms and is most conveniently expressed as a complex integral (in which i is the usual symbol for $\sqrt{-1}$):

$$f(t) = \frac{1}{\sqrt{2\pi}} \int\limits_{-\infty}^{\infty} a(\omega) \exp(i\omega t)\, d\omega.$$

The function $a(\omega)$, which may also be real, imaginary or complex, is obtained from $f(t)$ by an analogous integral:

$$a(\omega) = \frac{1}{\sqrt{2\pi}} \int\limits_{-\infty}^{\infty} f(t) \exp(-i\omega t)\, dt.$$

The factors of $1/\sqrt{(2\pi)}$ in these equations, not used by all authors, are introduced to maximise the symmetry between them.

If $f(t)$ is real, $a(\omega)$ obeys the relationship $a(-\omega) = a^*(\omega)$, where the * denotes the complex conjugate. In this case, all the information necessary to specify $f(t)$ is contained in the positive frequencies. An explicit representation in terms of sines and cosines is then possible:

$$f(t) = \int\limits_{0}^{\infty} S(\omega) \sin(\omega t)\, d\omega + \int\limits_{0}^{\infty} C(\omega) \cos(\omega t)\, d\omega.$$

The real functions $S(\omega)$ and $C(\omega)$ are related to the function $a(\omega)$ as follows:

$$S(\omega) = a^*(\omega) - a(\omega),$$

$$C(\omega) = a^*(\omega) + a(\omega).$$

The Fourier transform can also be applied in the spatial domain. If f is a function of a single spatial variable x (for example, altitude as a function of distance along a transect), the same equations are valid provided that t is replaced by x, and ω is replaced by q, corresponding to the angular spatial frequency (specified in radians per unit length). If f is a function of two Cartesian coordinates x and y, the appropriate integrals are most conveniently expressed in vector form:

$$f(\mathbf{x}) = \frac{1}{2\pi} \int\limits_{-\infty}^{\infty} \int\limits_{-\infty}^{\infty} a(\mathbf{q}) \exp(i\mathbf{q} \cdot \mathbf{x})\, dq_x\, dq_y,$$

$$a(\mathbf{q}) = \frac{1}{2\pi} \int\limits_{-\infty}^{\infty} \int\limits_{-\infty}^{\infty} f(\mathbf{x}) \exp(-i\mathbf{q} \cdot \mathbf{x})\, dx\, dy.$$

In these integrals, $\mathbf{q}$ is the vector (q_x, q_y) and $\mathbf{x}$ is the vector (x, y). Similar equations can be written for the three-dimensional case, in which case the factors before the integrals become $(2\pi)^{-3/2}$.

The integral expressions above assume that the variables are continuous. Fourier transforms can also be applied to discrete, regularly sampled data (for example an image in pixel form), in which case they are known as **discrete Fourier transforms** (DFTs). As an example, consider a function f of the single spatial variable x (the method is extended to two or three dimensions by analogy with the continuous Fourier transform). If the value of f is sampled at regular intervals Δx, such that f_j is its value at $x = j.\Delta x$ (where $j = 0$, $1, 2, \ldots, N - 1$), the DFT of f is the array a_k, defined as

$$a_k = \sum_{j=0}^{N-1} f_j \exp(-2\pi i j k / N)$$

for $k = 0, 1, 2, \ldots, N - 1$. The reverse transform is

$$f_j = \frac{1}{N} \sum_{k=0}^{N-1} a_k \exp(2\pi i j k / N).$$

The term with $k = 0$ corresponds to spatial frequency zero, as expected, and is in fact just equal to the sum of all the f_j. For $k \leq N/2$, the corresponding value of the spatial frequency q is $2\pi k / N \Delta x$. However, for $k \geq N/2$, the corresponding value of q is $2\pi (k - N) / N \Delta x$, i.e. negative. Thus, the DFT of an array sampled with a spacing Δx contains the spatial frequencies up to $\pm \pi / \Delta x$.

DFTs are usually implemented using an algorithm called the **fast Fourier transform**, which works best for values of N that are integer powers of 2.

Fourier transform spectrometry See *spectral resolution*.

FOV See *field of view*.

FPR (Flat-Plate Radiometer) U.S. thermal infrared radiometer, carried by *ESSA* (odd numbers), *ITOS* and *NOAA-1* satellites. Waveband: 7–30 μm.

Fragment-2 Russian optical/infrared mechanically scanned imaging radiometer, carried on *Meteor-Priroda-5* satellite. Wavebands: 0.4–1.1, 1.2–1.8, 2.1–2.4 μm. Spatial resolution: 80 m, 240 m, 480 m respectively. Swath width: 85 km.

Free space, impedance The *impedance* of free space (vacuum) to electromagnetic radiation. It is given by the expression

$$Z_0 = \sqrt{\frac{\mu_0}{\varepsilon_0}},$$

where μ_0 and ε_0 are the magnetic permeability and the electric permittivity, respectively, of free space. It has a value of approximately 377 Ω.

Frequency A measure of the repetition rate of any regularly repeating phenomenon, particularly the magnitude of the electric field at a point through which an electromagnetic *wave* is passing. If this variation obeys an equation of the form

$$E = A \sin(\omega t + \phi),$$

where E is the magnitude of the electric field, A (the amplitude), ω and ϕ (the phase) are constants and t is time, the wave is harmonic, possessing a unique frequency. The frequency f is specified in Hertz (cycles per second) and is given by

$$f = \frac{\omega}{2\pi},$$

where ω is the angular frequency, specified in radians per second (or in inverse seconds).

If the variation with time is not sinusoidal, it can be expressed as a (possibly infinite) sum of components with different values of A, ω and ϕ. The decomposition into frequency components, and the reconstitution of them into the original waveform, are performed mathematically using the *Fourier transform*.

Fresnel coefficients Coefficients relating the amplitudes of the reflected and transmitted components of an electromagnetic wave to the amplitude of the wave incident on a plane interface between two media. In their most general form, the coefficients are

$$r_{\text{perp}} = \frac{Z_2 \cos\theta_1 - Z_1 \cos\theta_2}{Z_2 \cos\theta_1 + Z_1 \cos\theta_2},$$

$$t_{\text{perp}} = \frac{2Z_2 \cos\theta_1}{Z_2 \cos\theta_1 + Z_1 \cos\theta_2},$$

$$r_{\text{par}} = \frac{Z_2 \cos\theta_2 - Z_1 \cos\theta_1}{Z_2 \cos\theta_2 + Z_1 \cos\theta_1},$$

$$t_{\text{par}} = \frac{2Z_2 \cos\theta_1}{Z_2 \cos\theta_2 + Z_1 \cos\theta_1},$$

where Z_1 and Z_2 are the *impedances* of the two media, and the wave is incident from medium 1 at an angle θ_1 to the normal to the interface. r and t are the amplitude coefficients for reflection and transmission, respectively, and the subscripts perp and par denote whether the electric field vector is oriented perpendicularly to, or parallel to, the plane containing the incident direction and the surface normal. The value of $\cos\theta_2$ is in general given by the expression

$$\cos\theta_2 = \sqrt{1 - \frac{\varepsilon_1 \sin^2\theta_1}{\varepsilon_2}},$$

where ε_1 and ε_2 are the *dielectric constants* of the two media. Note that, in general, Z_1, Z_2, ε_1, ε_2 and hence $\cos\theta_2$ may be complex.

If medium 1 is a vacuum (i.e. $\varepsilon_1 = 1$) and medium 2 is lossy but non-magnetic, the following fairly simple formulae exist for the power reflection coefficients:

$$|r_{\text{perp}}|^2 = \frac{(p - \cos\theta_1)^2 + q^2}{(p + \cos\theta_1)^2 + q^2},$$

$$|r_{\text{par}}|^2 = \frac{(\varepsilon_2' \cos\theta_1 - p)^2 + (\varepsilon_2'' \cos\theta_2 - q)^2}{(\varepsilon_2' \cos\theta_1 + p)^2 + (\varepsilon_2'' \cos\theta_2 + q)^2},$$

where

$$2p^2 = \sqrt{(\varepsilon_2' - \sin^2\theta_1)^2 + \varepsilon_2''^2} + \varepsilon_2' - \sin^2\theta_1$$

and

$$2q^2 = \sqrt{(\varepsilon_2' - \sin^2\theta_1)^2 + \varepsilon_2''^2} - \varepsilon_2' + \sin^2\theta_1.$$

The complex dielectric constant ε_2 of medium 2 is given by $\varepsilon_2' - i\varepsilon_2''$.

If both media are loss-free and non-magnetic, the amplitude coefficients become

$$r_{perp} = \frac{n_1\cos\theta_1 - n_2\cos\theta_2}{n_1\cos\theta_1 + n_2\cos\theta_2},$$

$$t_{perp} = \frac{2n_1\cos\theta_1}{n_1\cos\theta_1 + n_2\cos\theta_2},$$

$$r_{par} = \frac{n_1\cos\theta_2 - n_2\cos\theta_1}{n_1\cos\theta_2 + n_2\cos\theta_1},$$

$$t_{par} = \frac{2n_1\cos\theta_1}{n_1\cos\theta_2 + n_2\cos\theta_1},$$

where n_1 and n_2 are the (real) *refractive indices* of the two media. In this case, the expression for θ_2 simplifies to

$$\sin\theta_2 = \frac{n_1\sin\theta_1}{n_2}.$$

Note that if $n_1 > n_2$, it is possible for this formula to give a value of $\sin\theta_2$ which is greater than 1. If this condition holds, the magnitudes of the reflection coefficients are both 1, and the phenomenon is known as **total internal reflection**. The **critical angle** (smallest value of θ_1 for which this occurs) is given by $\sin^{-1}(n_2/n_1)$.

Note also that if $\theta_1 = \tan^{-1}(n_2/n_1)$, $r_{par} = 0$ and $t_{par} = n_1/n_2$. This value of θ_1 is called the **Brewster angle**.

Fronts, ocean See *ocean currents and fronts.*

Fully-developed sea See *significant wave height.*

Fuyo See *JERS.*

FY-1 ('Feng Yun' = 'wind and cloud' in Chinese) Chinese satellite operated by the state Meteorological and Oceanic Administrations of China. FY-1A was launched in September 1988 but failed after a month; FY-1B was launched in September 1990. Objectives: Operational meteorological observations. Orbit: Circular *Sun-synchronous LEO* at 890 km altitude. Period 102.9 minutes; inclination 99.0°; equator crossing time 07:45. Principal instruments: *VHRSR*.

FY-1 satellites are similar to the U.S. *NOAA* series. Further launches (FY-1C, FY-1D) are planned from 1998. These will carry the more advanced *MVISR* instrument.

FY-2 ('Feng Yun' = 'wind and cloud' in Chinese) Chinese satellite series, operated by the state Meteorological Administration of China. First satellite launched in

1992. Objectives: Operational meteorological observations. Orbit: *Geostation-ary* (105 °E). Principal instruments: *Scanning Radiometer*. The satellites also carry data collection packages (similar in operation to *Argos*).

URL: http://climate.gsfc.nasa.gov/~chesters/text/geonews.html#FENGYUN

G

GAC See *AVHRR*.

Gain The gain of an *antenna* that is transmitting radiation is defined as

$$G(\theta, \phi) = \eta_1 D(\theta, \phi),$$

where η_1 is the radiation efficiency of the antenna and D is its *directivity*. Unlike the directivity, the gain takes into account ohmic losses in the antenna.

Gas cell correlation radiometry See *spectral resolution*.

Gaussian stretch See *histogram matching*.

GCP See *ground control point*.

Gelbstoff See *ocean colour*.

Geobotany See *geology*.

Geocoding See *geometric correction*.

Geodetic Earth Observation Satellite-3 See *GEOS-3*.

Geodetic-geophysical satellite See *Geosat*.

Geoid A surface of constant gravitational potential lying close to the mean sea level. The geoid lies close to the *ellipsoid*, departures from this shape of the order of ± 100 m being due to variations in the density of the Earth's mantle and lithosphere, and to variations in the topography of the Earth's solid surface. Since the surface of a static fluid follows an equipotential surface, altimetric observations of the sea surface can be used to map the marine geoid provided the perturbing effects of currents, waves, tides and atmospheric pressure variations are removed (see *sea surface topography*).

Geoik (Space Geodetic Complex) Soviet/Russian satellites carrying *laser retroreflectors* which can be ranged from the ground for geodetic purposes. Launched

May 1988; 1989, 1990, 1994. <u>Orbit</u>: altitude 1500 km, inclination 74°, period 116 minutes.

Geology Remote sensing methods find widespread application in geology, for example in the identification and mapping of particular *rocks and minerals*, geomorphological and fault- and fold-zone mapping, determination of soil moisture, and monitoring erosion processes. Different geological units can be distinguished by a variety of methods including photographic interpretation, *classification* of multispectral imagery, *thermal inertia* mapping, *imaging radar* and *scatterometer* data. This information is often combined with data on *topography*, which can be obtained by *stereophotography*, stereo matching of *imaging radiometer* or imaging radar data, qualitative analysis of photography or imaging radiometer data acquired at low solar elevation angles, *interferometric SAR* or *laser profiler* data, to produce a geological interpretation of an area.

Discrimination between different rock types at visible and near infrared wavelengths is largely due to characteristic differences in spectral reflectance. Even the relatively broad spectral bands ($\approx$100 nm) provided by imaging radiometers such as *Landsat TM* permit the discrimination of a large number of rock types, although field-based validation is often required to make an interpretation. However, the development of *hyperspectral imagers* capable of resolving the finer structure in mineral absorption spectra promises significant enhancement to the number of rock types that can be identified. Optical/near infrared imagery can also be used for **geobotanical** analysis, i.e. the association of differences in plant growth with different geological conditions. The presence of toxic elements, especially copper, in the soil can retard or even prevent the growth of vegetation; certain indicator plants are associated with the presence of particular minerals; and excess of certain minerals can cause physiological changes manifested as identifiable alterations to tree structure or the yellowing (**chlorosis**) of leaves. These changes can be identified using the techniques of *vegetation mapping*.

At microwave frequencies, different geological units are distinguished largely on the basis of their surface roughness (see *Rayleigh criterion*) unless there are significant variations in surface moisture content. Multifrequency imagery, or data acquired over a range of incidence angles (notably scatterometer data), provide a significant improvement in the ability to identify different units. Analysis of image *texture*, especially the distribution of radar shadows and highlights, can also provide useful confirmation of an identification.

Lineaments and circular or elliptical structures, of geomorphic or tonal origin, can often be recognised in optical/near infrared or radar imagery, although their detectability is dependent on their orientation. In optical/near infrared imagery, linear features are less easily detected if they run parallel to the solar azimuth (the horizontal component of the direction of solar illumination), and if the imagery is scanned, features parallel to the scan lines are also hard to detect. In radar imagery, linear features are least detectable when they are aligned parallel to the *range direction*. However, such imagery does provide the advantage that lineaments can often be detected in the presence of forests or other vegetation. This is largely due to differences in vegetation growth, rather than to penetration of the radar signal through the vegetation canopy.

Geometric correction The process of mapping a remotely sensed image onto a chosen projection such as latitude/longitude or a coordinate grid. This involves the removal of essentially two types of geometric error: (1) systematic errors, arising from uncertainty in the position and viewing direction of the instrument and from any distortions inherent in the imaging process (for example, the skew introduced by mechanical scanning); and (2) random errors arising from geometrical irregularities during the imaging process (for example, non-uniform motion of the platform carrying the instrument). Some systematic errors can be corrected before the image is supplied to the user, in a process usually known as **geocoding**. Removal of any remaining geometric errors requires the use of *ground control points* to relate identifiable *pixels* to points on the Earth's surface having known coordinates. The image can either be corrected 'globally', defining a single coordinate transformation valid for all pixels and using the ground control points to obtain a least-squares estimate of the transformation's parameters, or it can be corrected 'locally', where different transformations are used in different parts of the image. In either case, the geometric correction involves *resampling* the pixels.

The relationship between the original *pixel* coordinates (x, y) and the transformed coordinates (u, v) in the new projection is specified by a pair of mapping functions:

$$u = f(x, y),$$
$$v = g(x, y),$$

and by an equivalent pair of inverse functions:

$$x = F(u, v),$$
$$y = G(u, v).$$

In practice these functions are often modelled as polynomials. The example below illustrates second-order polynomial functions: first-order polynomials can be obtained by setting a_3, a_4, a_5, b_3, b_4, and b_5 to zero.

$$u = a_0 + a_1 x + a_2 y + a_3 x^2 + a_4 y^2 + a_5 xy,$$
$$v = b_0 + b_1 x + b_2 y + b_3 x^2 + b_4 y^2 + b_5 xy.$$

A first-order polynomial (linear transformation) can correct for change of scale (including different scales in the x and y directions), rotation, and change of origin. A second-order polynomial (quadratic transformation) can also allow for skew effects.

Appropriate values of the coefficients a_0 to a_5 and b_1 to b_5 can be estimated from ground control points. The transformation can be determined globally, i.e. for the whole image, or piecewise over the image. Global transformations are normally preferred for satellite images since the errors to be corrected are usually constant over the image. Once the transformation has been determined, the digital numbers of pixels in the original image are assigned to pixels in the new (transformed) image by resampling.

Geometric optics model See *Kirchhoff model*.

GEOS-3 (Geodynamics Experimental Ocean Satellite) U.S. satellite, operated by *NASA*, launched April 1975, terminated December 1978. Objectives: Dedicated radar altimeter mission for topographic mapping of oceans. Orbit: Circular *LEO* at 843 km altitude. Period 101.8 minutes; inclination 115°. Principal instruments: *Radar altimeter*. GEOS-3 also carried a laser retroreflector and dual-frequency radio transmitter, for precise determination of its position.

Geosat (Geodetic-geophysical satellite) U.S. military/civilian satellite operated by the U.S. Navy and by *NOAA*. Launched March 1985, terminated January 1990. Objectives: Dedicated radar altimetry mission for geodetic mapping at high and intermediate spatial resolution, ocean and land-ice topography. Orbit: Circular *LEO* at 800 km (for GM) and 793 km (for ERM) altitudes. Period 101 minutes; inclination 108°. ERM orbit was an *exactly repeating orbit* (244 orbits in 17 days). Principal instruments: *Radar altimeter*.

The first 18-month period of the mission was designated the 'geodetic mission' (GM), providing classified geodetic data of high spatial resolution. The remainder of the mission was the 'exact repeat mission' (ERM), which duplicated the orbit of *Seasat*. The data from the ERM are unclassified.

See also *GFO*.

URL: http://leonardo.jpl.nasa.gov/msl/QuickLooks/geosatQL.html

Geoscience Laser Altimeter System See *GLAS*.

Geostationary See *geosynchronous orbit*.

Geostationary Meteorological Satellite See *GMS*.

Geostationary Operational Environmental Satellite See *GOES*.

Geostationary Orbit Meteorological Satellite See *GOMS*.

Geosynchronous orbit A *prograde* satellite *orbit* with a *nodal period* of one *sidereal day* (i.e. equal to the period of the Earth's rotation about its axis). Such a satellite will be in exactly the same position with respect to the Earth's surface at intervals of one sidereal day. If the *inclination* and *eccentricity* of the orbit are both zero, it is said to be **geostationary**, and the satellite remains in a fixed position relative to the Earth's surface. A geostationary orbit has a *semimajor axis* of 42 170 km, and so is located at a height of approximately 35 800 km above the equator. Geostationary orbits are used for some meteorological satellites, as well as for data relay, telecommunications and television transmission satellites.

The view of the Earth's surface from a geostationary satellite is illustrated in the figure, which shows a latitude/longitude grid (at 10° intervals) with the correct perspective. The black circles are labelled with the elevation angle, in degrees, of the line of sight to the satellite. In practice, useful surface data cannot be acquired for elevation angles below about 20°, because of the long path length through the atmosphere and the distortion produced by the oblique viewing geometry. This gives an effective coverage between latitudes 60° N and 60° S at the same longitude as the satellite, equivalent to about one quarter of the Earth's surface.

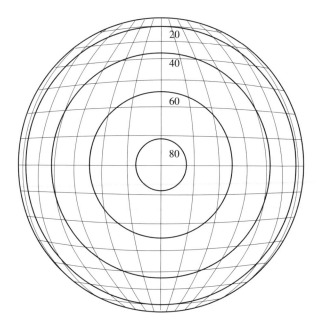

The following table summarises the geostationary meteorological satellites used for remote sensing.

Satellite	Longitude (°, East positive)	Operated by
GOES-West	−135	U.S.A.
GOES-East	−75	U.S.A.
Meteosat	0	ESA/EUMETSAT
Insat	74	India
GOMS/Elektro	76	Russia
FY-2	105	China
GMS/Himawari	140	Japan

GFO (Geosat Follow-On) U.S. satellite, operated by the U.S. Navy, launched February 1998. Objectives: Provision of operational *radar altimeter* data. Orbit: Circular *LEO* (not Sun-synchronous) at 800 km altitude; inclination 108°. Period: 100 minutes. Principal instruments: *Radar Altimeter, WVR*. The orbit of GFO duplicates the ERM of the *Geosat* mission.

URL: http://gfo.bmpcoe.org/Gfo/default.htm

GFZ-1 German satellite carrying *laser retroreflectors* which can be ranged from the ground for geodetic purposes. Launched April 1995. Orbit: altitude 400 km initially, declining to about 150 km over 3 years, inclination 51.6°.

Glaciers A glacier is a persistent accumulation of snowfall occupying an area of land of at least 0.1 km^2, moving in response to gravitational forces. The *ice sheets* of Antarctica and Greenland are glaciers, as are the smaller ice caps with areas typically between 1000 and 10 000 km^2, but the term is often used only for the smaller mountain glaciers.

Glaciers play a major role in the global hydrological cycle. They cover an area of approximately 15.8 × 10^6 km^2 (11% of the Earth's surface) and contain approximately 3.30 × 10^{16} m^3 of ice, representing over 80% of the Earth's fresh water and 2% of the total water. Partial melting of the Greenland Ice Sheet is estimated to have contributed 25 mm to the 120 mm rise in sea level that has occurred over the last century – the Antarctic Ice Sheet has probably contributed zero to this total, but the world's smaller glaciers are thought to have contributed about 50 mm. At the last ice-age maximum, 18 000 years ago, the volume of the world's glaciers was approximately 2.5 times as great as it is now, covering three times the present area and lowering the global sea level by about 120 m. At the last interglacial minimum, about 12 000 years ago, the global sea level was approximately 6 m higher than at present, mostly as a result of the almost total melting of the Greenland Ice Sheet. Glaciers also exert a significant influence on the global climate, particularly through *albedo* feedback (in the case of the ice sheets) on the *Earth radiation budget*. Monitoring of the response of smaller glaciers to climate variations represents an important and sensitive method of studying global climate change.

In some parts of the world, glaciers represent water resources. They can also pose hazards through damage to fixed facilities as a result of glacier movement, the sudden discharge of large quantities of meltwater, and the discharge of icebergs into seas, lakes or rivers.

Snow is initially deposited on the surface of a glacier at low density, typically 50 to 100 kg m^{-3}. The density increases as a result of wind-packing, freeze-thaw cycling, and compression by overlying snow, and at sufficient depth the density approaches the value of 917 kg m^{-3} characteristic of pure ice. The input of material to a glacier is termed **accumulation**. Loss of material is termed **ablation**, and is caused by runoff of meltwater, evaporation, removal by wind, and calving of icebergs. The accumulation and ablation rates vary with season and with position on the glacier; the **equilibrium line** separates the accumulation area, for which the annual balance is positive, from the ablation area, for which it is negative. Interannual variations in the position of the equilibrium line are strongly correlated with variations in the total mass balance of a glacier.

Remote sensing methods can be used to study many of the properties of glaciers. Spatial extent and surface features can be mapped using visible-wavelength or *synthetic aperture radar* imagery. Comparison of the positions of identifiable surface features, such as crevasses, in time-series of images, can be used to determine surface velocities. Surface topography can be measured using *radar altimeter*, *interferometric SAR* or *laser profiler* measurements and can also be deduced from visible-wavelength imagery using shape-from-shading methods. Optical/near infrared imagery delineates the **snowline** (the boundary between snow-covered and snow-free parts of the glacier), and the position of the snowline at the time of maximum melt normally corresponds to the equilibrium line. More detailed subdivision of the surface of a glacier, into dry snow (negligible melting), percolation (localised summer melting followed by refreezing), wet snow (thorough wetting of the surface snow), superimposed ice (where so much meltwater is produced during the summer that the resultant ice layers merge into a continuous mass) and bare ice zones, is the subject of current research using optical/near infrared and *synthetic aperture radar* data. As with the snowline, monitoring of interannual changes in the boundaries between these zones should give improved accuracy in knowledge of the glacier's mass balance.

GLAS (Geoscience Laser Altimeter System) U.S. backscatter lidar/laser profiler for cloud and aerosol profiling, planned for inclusion on *EOS-Laser Alt* satellites. Wavelength: 532 nm (Nd-YAG laser), 1063 nm (for laser profiler). Spatial resolution: 75–200 m (horizontal), 50–150 m (vertical – lidar), 0.1 m (vertical – laser profiler). Pulse repetition frequency: 40 Hz (sampling interval 188 m).

The GLAS instrument was formerly known as **GLRS** (Geoscience Laser Ranging System).

GLCM See *grey level co-occurrence matrix*.

GLI (Global Imager) Japanese ultraviolet/optical/infrared imaging radiometer, planned for inclusion on *ADEOS II* satellite. Wavebands: 23 bands (width 10 to 15 nm) between 0.38 and 0.83 μm, 6 bands between 1.05 and 2.22 μm, 7 bands between 3.72 and 11.95 μm. Spatial resolution: 0.25–1 km at nadir. Swath width: 2000 km.

The design of the *GLI* is based on that of the *OCTS* instrument on *ADEOS I*.

URL: http://hdsn.eoc.nasda.go.jp/guide/guide/satellite/sendata/gli_e.html

Global Area Coverage See *AVHRR*.

Global Imager See *GLI*.

Global Ozone Monitoring by Occultation of Stars See *GOMOS*.

Global Ozone Monitoring Experiment See *GOME*.

Global Positioning System See *GPS*.

GLONASS (Global Orbiting and Navigation Satellite System) Russian satellite navigation system, similar to *GPS* though using slightly different orbital and data transmission parameters. The first GLONASS satellite was launched in October 1982, and the system became fully operational (16 operational satellites) from 1996.

URL: http://www.rssi.ru/SFCSIC/glonass.html

GLRS See *GLAS*.

GMS (Geostationary Meteorological Satellite) Japanese satellites operated by the Japanese Meteorological Agency. GMS-1 was launched in July 1977; further satellites have been launched to maintain continuity of observation. Objectives: Operational meteorology, particularly through observations of cloud cover and dynamics. Orbit: *Geostationary* (longitude 120°, 140°). Principal instruments: *VISSR*.

GMS is also known as 'Himawari' (Japanese for 'sunflower'). The first satellites were based on the *GOES* satellites but an indigenous design has been evolved. The satellites also carry a data collection package (similar in function to *Argos*) and a solar–terrestrial physics package.

URL: http://www.jwa.go.jp/gms.html

GOES (Geostationary Operational Environmental Satellite) Series of U.S. satellites operated by *NOAA* and *NASA*, also known as **SMS** (synchronous meteorological satellites). The GOES programme began in May 1974. Two satellites (GOES-East and GOES-West) are in operation at any time, with nominal lifetimes of 5 years. Objectives: Operational meteorological observations, collection and relay of weather data. Orbit: *Geostationary* (longitude 75° W [GOES-East], 135° W [GOES-West]). Principal instruments: *VISSR* (to 1978), *VAS* (from 1980).

The GOES satellites also carry a data collection package (similar in function to *Argos*), and search and rescue and solar–terrestrial physics packages. See also *GOES-Next*.

URL: http://goeshp.wwb.noaa.gov/
http://www.goes.noaa.gov

GOES Imager U.S. optical/infrared spin-scan imaging radiometer, carried on *GOES-Next* satellites. Wavebands: 0.55–0.75, 3.8–4.0, 6.5–7.0, 10.2–11.2, 11.5–12.5 µm. Spatial resolution: 1 km (optical) at nadir, 4–8 km (infrared) at nadir. Field of view: Full Earth disc as seen from *geostationary* orbit. Scan time: 25 minutes (full disc), 3 minutes (3000 km × 3000 km), 40 seconds (1000 km × 1000 km).

GOES-Next (Geostationary Operational Environmental Satellite – next (or second) generation) Series of U.S. satellites operated by *NOAA* and *NASA*. GOES-8, the first satellite of this series (which is the successor to *GOES*), was launched in April 1994. Objectives: Operational meteorological observations, collection and relay of weather data. Orbit: *Geostationary* (longitude 75° W [GOES-East], 135° W [GOES-West]). Principal instruments: *GOES Imager, GOES Sounder*. Also carries data collection and solar terrestrial physics packages.

URL: http://climate.gsfc.nasa.gov/~chesters/goesproject.html

GOES Sounder U.S. infrared spin-scan radiometer, carried on *GOES-Next* satellites. Wavebands: 18 channels between 3.7 and 14.7 mm. Spatial resolution: 8 km at nadir. Field of view: Full Earth disc as seen from *geostationary* orbit. Scan time: 464 minutes (full disc), 42 minutes (3000 km × 3000 km), 5 minutes (1000 km × 1000 km).

The instrument will provide profiles of atmospheric temperature and water vapour, as well as measurements of *SST* and total atmospheric ozone.

GOME (Global Ozone Monitoring Experiment) European nadir-looking ultraviolet/optical/near infrared spectrometer, carried on *ERS-2*. Wavebands: 3584 channels between 240 and 790 nm. Spectral resolution: 0.2–0.4 nm. Spatial resolution: 40 km × 40 km to 40 km × 320 km. Height resolution: 1 km.

The instrument views three strips parallel to the along-track direction, and also views the Sun and Moon for calibration. Ozone concentration profiles are retrieved from backscatter and differential absorption measurements. A modified version of GOME, *OMI*, is planned for inclusion on *Metop-1*.

URL: http://earth1.esrin.esa.it/f/eeo2.402/eeo4.108

GOMOS (Global Ozone Monitoring by Occultation of Stars) European star-viewing ultraviolet/optical/near infrared limb sounder, for profiling of atmospheric temperature, aerosols, H_2O, O_3, NO_2 and other gases, planned for inclusion on *Envisat*. Waveband: 0.25–0.95 μm. Spectral resolution: 1.2 nm (0.25–0.68 μm); 0.18 nm (0.756–0.773 and 0.926–0.952 μm). Height resolution: 1.7 km. Height range: 20–100 km.

GOMOS will measure absorption spectra using the light from about 30 stars per orbit, giving improved spatial coverage compared with Sun-viewing and Moon-viewing limb sounders. Light is dispersed by a diffraction grating and detected by a two-dimensional CCD.

URL: http://envisat.estec.esa.nl/instruments/gomos/

GOMS (Geostationary Operational Meteorological Satellite) Russian satellite series providing a similar service to *GOES* satellites. First satellite launched November 1994, with a nominal lifetime of 3 years. The system became partially operational in June 1996. Objectives: Operational meteorological observations (particularly cloud cover and dynamics). Orbit: *Geostationary* (longitude 76° E). Principal instruments: *STR*. GOMS-1 also carries a data collection package (similar in function to *Argos*) and a solar–terrestrial physics package.

GOMS-1 is also known as **Elektro**.

URL: http://sputnik.infospace.ru/goms/engl/goms_e.htm

GPS (Global Positioning System) U.S. satellite navigation system, consisting of 21 operational NAVSTAR satellites in near-circular orbits in six orbital planes. The orbits are chosen to ensure that at least four satellites are above the horizon at any time for any observer. The satellites are at a nominal altitude of 20 200 km (orbital period 718 minutes). Early satellites had an inclination of 63°; this has now been reduced to 55°. The system uses one-way (downlink) coded transmissions, at 1228 and 1572 MHz, with very accurately controlled transmission times. GPS receivers use timing and ephemeris data transmitted by the satellites to determine position, with a nominal accuracy of 100 m (civilian users), 18 m (military users). Higher accuracies are possible for relative position measurements, since the position errors are strongly spatially correlated.

See also *GLONASS*.

URL: http://wwwhost.cc.utexas.edu/ftp/pub/grg/gcraft/notes/gps/gps.html

Greben Russian nadir-viewing radar altimeter carried on *Mir-1*. Frequency: K_u band (13.8 GHz). Pulse length (uncompressed): 1.7 μs; (compressed): 12.5 ns. Range precision: 10 cm. Beam-limited footprint: 13 km. Pulse-limited footprint: 2.3 km.

Greenness See *tasselled-cap transformation*.

Grey level See *digital number*.

Grey level co-occurrence matrix A statistical technique for quantifying image *texture*. If the digital numbers in a single band of an image are drawn from a set of N integers (e.g. $N = 64$ for 6-bit data), the grey level co-occurrence matrix (GLCM, also called the **spatial dependency matrix**) is an $N \times N$ square matrix. The matrix is evaluated within a window of the image as follows. First, all the elements P_{ij} of the GLCM are set to zero and a vector displacement (x, y) is chosen. Second, each pixel in the image window is examined in turn. If the pixel has coordinates (x_p, y_p) and digital number i, and the pixel at coordinates $(x_p + x, y_p + y)$ has digital number j, the value of P_{ij} is increased by one. When this process is completed, the element P_{ij} of the GLCM shows the frequency with which digital number j occurs with the chosen displacement from digital number j.

Various parameters can be calculated from the GLCM, and these are often used as statistical measures of the image texture within the window. Definitions of the commonest of these are given below:

$$\sum P_{ij}^2 \qquad \text{energy}$$

$$\sum P_{ij} \ln P_{ij} \qquad \text{entropy}$$

$$\sum |i - j|^a P_{ij}^b \qquad \text{contrast}$$

The sums are taken over all values of i and j from 1 to N. The energy (also called the angular second moment) is a measure of homogeneity. The contrast is a measure of local variability. Common values of a and b in the definition of contrast are 2 and 1, respectively.

Grid format See *raster format*.

Ground control point (GCP) A feature recognisable both in an image and on the ground (or on a map), used for *geometric correction* of an image. Suitable GCPs must have well-defined positions in the image, usually to the nearest pixel, and well-defined positions on the ground, again usually to an accuracy corresponding to one pixel. Commonly used GCPs for optical and infrared-wavelength imagery include road intersections, field boundaries, edges of water bodies (including coastline features) etc. For radar imagery, suitable GCPs may not exist, in which case it may be necessary to emplace *radar transponders* as artificial GCPs.

The position of a GCP on the ground can be determined by surveying (including the use of *GPS* surveying methods), or by measurement from a map. The absolute minimum number of GCPs that are needed to perform a geometric correction is determined by the number of parameters in the model defining the geometric correction to be applied, but in practice typically 10 times this number will be used and a statistical (least-squares) fit made to the model. One normally tries to arrange that most of the GCPs are distributed around the edge of the image, with a few in the middle. The usual criterion of a successful fit to the model is that the root mean square error for all the GCPs should be less than one pixel.

If an image is to be registered to another image, rather than to a particular map projection, it is only necessary that the GCPs be identifiable, and have well-defined positions, in both images.

Ground segment The terrestrial facilities, operated by space agencies etc., necessary to maintain the provision of spaceborne remote sensing data. The usual elements of a ground segment are a mission control centre, which issues commands to the spacecraft and analyses 'housekeeping' data transmitted from it, one or more *receiving stations* that receive the remotely sensed data either directly from the spacecraft or from a *relay satellite*, and data processing, archiving and dissemination facilities. The various elements of a ground segment may be geographically distributed. The spacecraft itself, and any relay satellites that may be in use, constitute the **space segment**.

Group velocity See *dispersion*.

Hadamard transform A transformation from the spatial (image) domain to the spatial frequency domain, or vice versa. The Hadamard transform is thus similar to the *Fourier transform*, but whereas the latter analyses a spatial variable in terms of sine and cosine functions, the former uses **Walsh functions**. Since these take only values of ±1, they are better suited to the analysis of digital image data. They are also readily computed.

The Hadamard transform $\mathbf{F}$ of an image $\mathbf{f}$ is defined by the operation

$$\mathbf{F} = \mathbf{P}\mathbf{f}\mathbf{Q},$$

where $\mathbf{P}$ and $\mathbf{Q}$ are Hadamard matrices, and $\mathbf{f}$ and $\mathbf{F}$ are treated as matrices. The 2×2 Hadamard matrix is

$$\mathbf{P}_2 = \begin{bmatrix} 1 & 1 \\ 1 & -1 \end{bmatrix}.$$

Hadamard matrices of any order that is a power of 2 are defined by

$$\mathbf{P}_{2k} = \begin{bmatrix} \mathbf{P}_k & \mathbf{P}_k \\ \mathbf{P}_k & -\mathbf{P}_k \end{bmatrix}.$$

Thus, for example, the Hadamard matrix of order 8 is

$$\mathbf{P}_8 = \begin{bmatrix} 1 & 1 & 1 & 1 & 1 & 1 & 1 & 1 \\ 1 & -1 & 1 & -1 & 1 & -1 & 1 & -1 \\ 1 & 1 & -1 & -1 & 1 & 1 & -1 & -1 \\ 1 & -1 & -1 & 1 & 1 & -1 & -1 & 1 \\ 1 & 1 & 1 & 1 & -1 & -1 & -1 & -1 \\ 1 & -1 & 1 & -1 & -1 & 1 & -1 & 1 \\ 1 & 1 & -1 & -1 & -1 & -1 & 1 & 1 \\ 1 & -1 & -1 & 1 & -1 & 1 & 1 & -1 \end{bmatrix}.$$

HALOE (Halogen Occultation Experiment) U.S. infrared limb-sounder (Sun-viewing), carried on *UARS* satellite. Waveband: 2.43–10.25 μm (gas filter correlation spectrometer). Spatial resolution: 6.2 km horizontal, 1.6 km vertical.

Absorption measurements are used to obtain profiles of HF, HCl and other minor stratospheric constituents.

URL: http://haloedata.larc.nasa.gov/home.html

Halogen Occultation Experiment See *HALOE*.

Haute Resolution Visible See *HRV*.

Haze General term for sub-micron atmospheric constituents (principally molecules and *aerosols*) causing appreciable *scattering* of radiation, especially in the ultraviolet and visible parts of the electromagnetic spectrum. See *atmospheric correction*.

HCMM (Heat Capacity Mapping Mission) U.S. satellite, launched April 1978, operated to August 1980. Objectives: Experimental mission to measure *thermal inertia*, surface temperature and heat budget. Orbit: Circular *Sun-synchronous LEO* at 620 km altitude (540 km after February 1980). Period 97 minutes (96 after Feb 1980); inclination 98°; equator crossing time 14:00 (ascending node). *Exactly repeating orbit*. Principal instruments: *HCMR*.
Also known as AEM-1 (Applications Explorer Mission).

HCMR (Heat Capacity Mapping Radiometer) U.S. optical/infrared imaging radiometer, carried on *HCMM* satellite. Wavebands: 0.55–1.11 μm, 10.5–12.5 μm. Spatial resolution: 500 m (600 m for thermal infrared band). Swath width: 900 km.

Heat Capacity Mapping Mission See *HCMM*.

Heat Capacity Mapping Radiometer See *HCMR*.

Height, orbital The altitude above the Earth's surface of a satellite in a circular *orbit* is not constant, because the Earth is not spherical (see *ellipsoid*). It is conventional to define the orbital height as the difference between the orbital radius and the length of the Earth's equatorial axis.

Hemispherical albedo See *bidirectional reflectance distribution function*.

HH-polarisation A term used in *radar*, to describe a signal that is both transmitted and received in *horizontal polarisation*. This is a *co-polarised* mode of operation.

High-boost filter See *sharpening*.

Highlighting, radar Enhanced *backscatter* in a *radar* image of rough terrain, arising from areas in which the surface normal is close to the radar's line of sight. See *side-looking radar*.

High-pass filter See *edge detection*.

High Resolution CCD camera See *HRCC*.

High Resolution Doppler Imager See *HRDI*.

High Resolution Dynamics Limb Sounder See *HiRDLS*.

High Resolution Geometry See *HRG*.

High Resolution Infrared Radiometer See *HRIR*.

High Resolution Infrared Sounder See *HIRS*.

High Resolution Picture Transmission See *AVHRR*.

High Resolution Visible See *HRV*.

High Resolution Visible Infrared See *HRV*.

Himawari See *GMS*.

HiRDLS (High Resolution Dynamics Limb Sounder) U.S./U.K. thermal infrared (emission) limb sounder, for global profiling of atmospheric H_2O, O_3, NO_2, CH_4 and other gases, aerosols, and measurement of cloud-top temperatures, planned for inclusion on *EOS-Chem* satellites. Waveband: 21 channels from 6.12–17.76 µm. Spatial resolution: 400 km (along-track), 10 km (across-track), 1 km (vertical). Swath width: 3000 km.

URL: http://eos.acd.ucar.edu/hirdls/home.html

HIROS See *ALOS*.

HIRS (High Resolution Infrared Sounder) U.S. optical/infrared scanning radiometer, carried on *Nimbus-6* satellite. Wavebands: 17 channels between 0.69 and 17 µm. The instrument was used to provide temperature and water vapour profiles for the atmosphere.

HIRS/2 (High Resolution Infrared Sounder-2) U.S. optical/infrared scanning radiometer, carried on *NOAA-6* onwards and *TIROS-N*, as part of the *TOVS* package. Wavebands: 12 channels between 6.72 and 14.95 µm, 7 channels between 3.76 and 4.57 µm, single channel at 0.69 µm. Spatial resolution: 20 km at nadir. Swath width: 2240 km. Altitude range: Up to 40 km.

   Like its predecessor *HIRS*, the instrument measures profiles of atmospheric temperature and water vapour.

HIRS/3 (High Resolution Infrared Sounder-3) U.S. optical/infrared scanning radiometer, planned for inclusion on forthcoming NOAA satellites and on *Metop-1*. Wavebands: 0.69, 3.76, 4.00, 4.33, 4.45, 4.47, 4.52, 4.57, 6.52, 7.33, 9.71, 11.11, 12.47, 13.35, 13.64, 13.97, 14.22, 14.49, 14.71, 14.95 µm. Spatial resolution: 20 km at nadir. Swath width: 1000 km.

Histogram The histogram of a single-band image is a graph showing the total number of pixels $h(x)$ with *digital number* x. It is a discrete representation of the probability distribution function of the image brightness. Separate histograms can be constructed for each band of a multi-band image: histograms can also be constructed for regions of an image.

Histogram matching A form of *contrast enhancement* in which the *digital numbers* of the pixels in an image are reassigned so that the image *histogram* has a specified form.

If image 1 has a total of N_1 pixels, each of which can have a digital number between 0 and n_1, and an image histogram $h_1(x)$, and image 2 has the corresponding values N_2, n_2, $h_2(x)$, the transform $f(x)$ that, when applied to image 1, will match its histogram to that of image 2, is given by

$$f(x) = g_2^{-1}(g_1(x)),$$

where

$$g_1(x) = \frac{n_1}{N_1} \sum_{i=0}^{x} h_1(i),$$

$$g_2(x) = \frac{n_2}{N_2} \sum_{i=0}^{x} h_2(i),$$

and $g_2^{-1}(x)$ is the inverse function of $g_2(x)$, defined such that

$$g_2^{-1}(g_2(x)) = g_2(g_2^{-1}(x)) = x.$$

In general, quantisation (digitisation) errors will mean that the function $f(x)$ cannot be performed exactly.

The simplest type of histogram matching is **histogram equalisation**, in which the object is to find a transform $f(x)$ that will cause image 1 to have a 'flat' histogram, i.e. all digital numbers occurring with equal frequency. In this case we can set $h_2(x) = N_1/(n_1 + 1)$, giving

$$f(x) = \frac{n_1 + 1}{N_1} \sum_{i=0}^{x} h_1(i) - 1.$$

Histogram equalisation is often performed automatically by image processing systems as a method of improving the image contrast. A similar automatic histogram-matching process is the **Gaussian stretch**, in which the image is transformed so that it has a Gaussian histogram with a defined mean and standard deviation.

Image-to-image histogram matching is often used when images are to be joined together in a *mosaic*, in order to minimise any abrupt changes in image contrast at the boundaries between the images.

Horizontal polarisation A term used in *side-looking radar* and *synthetic aperture radar* to describe linearly polarised radiation in which the electric field vector is horizontal. Compare *vertical polarisation*.

Hotelling transformation See *principal components*.

Hough transform A technique in image processing for detecting straight lines in images. If an image pixel is represented by its Cartesian coordinates (x, y), and the Hough transform space is represented by the parameters r and θ, the transformation operates as follows. For every pixel (x, y) to be considered, calculate

$$r = x \cos \theta + y \sin \theta$$

for all values of θ, and increment the value stored in the Hough transform at (r, θ) by 1. When this process has been completed, peaks in the Hough transform array correspond to the presence of lines in the image. The values of r and θ give the position and orientation of the line, as shown in the figure, and the height of the peak corresponds to the length of the line. The transform does not give the absolute position of the line.

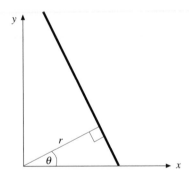

The Hough transform is usually applied after *edge detection* or image thresholding, so that only certain pixels in the image are considered.

HPBW Half-power *beamwidth*.

HRCC (High Resolution CCD camera) Brazilian optical/near infrared *CCD* imaging radiometer to be carried on the *CBERS* satellite. Wavebands: 0.45–0.52, 0.51–0.73, 0.52–0.59, 0.63–0.69, 0.77–0.89 µm. Spatial resolution: 20 m. Swath width: 120 km, steerable to ±32° from the sub-satellite track.

HRDI (High Resolution Doppler Imager) U.S. instrument for measuring wind speed in the upper troposphere and stratosphere, carried on *UARS*. HRDI is a nadir-viewing optical/near infrared Fabry–Perot interferometer designed to measure Doppler shifts in O_2 absorption lines. Waveband: Channels at 558, 630, 631, 687, 688, 692, 693, 724, 761, 764, 765, 767, 776 nm. Spectral resolution: 0.002 nm tunable within $\approx$1 nm range in each channel. Accuracy: $5 \, \mathrm{m \, s^{-1}}$ (troposphere and stratosphere), $15 \, \mathrm{m \, s^{-1}}$ (mesosphere, thermosphere).

URL: http://eosdata.gsfc.nasa.gov:80/DATASET_DOCS/HRDI_dataset.html

HRG (High Resolution Geometry) Optical/near infrared imager planned for inclusion on *SPOT-5*. The HRG is based on *HRV* and *HRVIR* but provides higher spatial resolution. Wavebands: 1: 0.50–0.59, 2: 0.61–0.68, 3: 0.78–0.89, 4: 1.58–1.75, panchromatic: 0.51–0.73 µm.

SPOT-5 will carry three HRG instruments, viewing at nadir and at 19.2° forward and aft of nadir. In addition, the instruments can be pointed at up to 27° either side of nadir.

The HRG instruments can be operated in three modes: X (multispectral), S (stereo) and H (high resolution), with spatial resolutions (in metres) as shown below. — denotes that data from a band are not available in that mode.

Mode	Band 1	2	3	4	pan
X	10	10	10	20	–
S	–	–	10	–	5
H	10	10	10	–	5

Swath width: 60 km (nadir viewing); 80 km (off-nadir). The swath width in X and H modes can be increased to 240 km by using all three instruments.

HRIR (High Resolution Infrared Radiometer) U.S. infrared mechanically scanned imaging radiometer, carried on *Nimbus-1* and *-2* satellites. Waveband: 3.4–4.2 µm. The Nimbus-2 instrument also had a near infrared band at 0.7–1.3 µm. Spatial resolution: 3 km at nadir. Swath width: 2700 km.

HRPT See *AVHRR*.

HRV (Haute Resolution Visible, or High Resolution Visible Sensor) French optical/near infrared *CCD* (*pushbroom*) imaging radiometer carried on *SPOT-1* to *-3*. Wavebands: multispectral (XS) mode: 0.50–0.59, 0.61–0.68, 0.79–0.89 µm; panchromatic mode: 0.51–0.73 µm. Spatial resolution: 20 m (XS), 10 m (panchromatic). Swath width: Two sensors each with a 60 km swath, overlapping by 3 km. Steerable to 27° either side of the sub-satellite track. This allows for stereo imaging and also for more frequent observation of a particular location.

URL: http://www.spotimage.fr/anglaise/system/satel/ss_paylo.htm

HRVIR (High Resolution Visible and Infrared Sensor) French optical/near infrared *CCD* (*pushbroom*) imaging radiometer carried on *SPOT-4*. Wavebands: 1: 0.50–0.59; 2: 0.61–0.68; 3: 0.79–0.89; 4: 1.58–1.75 µm. Spatial resolution: 10 m or 20 m. Swath width: Two sensors each with a 60 km swath, overlapping by 3 km. Steerable to 27° either side of the sub-satellite track. This allows for stereo imaging and also for more frequent observation of a particular location. The HRVIR is based on the *HRV* instrument.

URL: http://www.spot.com/anglaise/spot4/Aperfo.HTM

HSI (Hyperspectral Imager) U.S. optical/near infrared imaging spectrometer, carried on *Lewis*. Wavebands: Panchromatic band: 0.48–0.75 μm; 128 channels between 0.40 and 1.0 μm; 256 channels between 0.90 and 2.5 μm. Spectral resolution: 5 nm (0.4–1.0 μm); 6.4 nm (0.9–2.5 μm). Spatial resolution: 5 m (panchromatic); 30 m (hyperspectral). Swath width: 12.9 km (panchromatic); 7.7 km (hyperspectral).

Hue See *IHS display*.

Huffman code A common form of data *compression* for digital data, making use of the fact that some data values are likely to be more common than others. The digital numbers representing the data are replaced by codes consisting of sequences of 0s and 1s, such that frequently occurring digital numbers are assigned short codes and infrequently occurring numbers are assigned long codes. The codes themselves are not strictly binary numbers, since any leading zeros are significant. The compression that can be achieved using a suitable Huffman code usually approaches fairly closely to the theoretical *information content* of the data.

Humidity See *water vapour*.

HV-polarisation A term used in *radar*, to describe a signal that is transmitted in *horizontal polarisation* and received in *vertical polarisation*. This is a *cross-polarised* mode of operation.

Hybrid classification An image *classification* technique combining *unsupervised classification* and *supervised classification*. A typical hybrid classification algorithm involves the following steps: unsupervised classification (usually of a representative subset of the whole image) to determine the number of distinguishable information classes (clusters) present in the image; comparison of the distinguishable information classes with the desired ground-cover classes; modification of ground-cover classes to take account of distinguishability; splitting, merging and deletion of clusters to correspond to the modified ground-cover classes; supervised classification of the whole image. The process is normally an iterative one.

Hyperspectral Imager See *HSI*.

IASI (Improved Atmospheric Sounder Interferometer) French nadir-viewing infra-red spectrometer, planned for inclusion on *Metop*. <u>Waveband</u>: 3.62–5.00, 5.00–8.26, 8.26–15.5 μm. <u>Spectral resolution</u>: 45, 30, 15 GHz respectively (Fourier transform interferometer). <u>Spatial resolution</u>: 15 km.

IASI is intended to measure profiles of temperature, water vapour, ozone, methane and other constituents of the troposphere.

Ice, dielectric properties

Optical/infrared region
The graph shows the dielectric constant of pure ice as a function of wavelength between 0.9 μm and 50 μm.

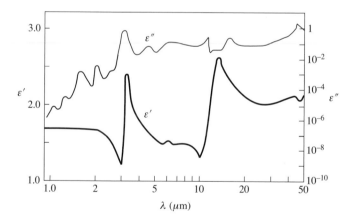

The *emissivity* of ice is approximately 0.96 in the range 8–14 μm.

Microwave region
The real part ε' of the dielectric constant of pure ice is about 3.17 for frequencies between 10 MHz and 1000 GHz. The imaginary part ε'' is dependent on the temperature and frequency, with a minimum value at about 3 GHz. The following table gives some approximate values of ε''.

f (GHz)	$-5\,^{\circ}\mathrm{C}$	$-15\,^{\circ}\mathrm{C}$
1	0.0006	0.0004
2	0.0004	0.0002
5	0.0004	0.0003
10	0.0008	0.0005
20	0.002	0.001
50	0.005	0.004
100	0.01	0.009

See also *sea ice*.

Ice sheets The large ice sheets of Antarctica and Greenland together cover an area of $15.3 \times 10^6\,\mathrm{km}^2$ and contain $32.7 \times 10^6\,\mathrm{km}^3$ of ice, representing over 80% of the Earth's fresh water. They play an important yet imperfectly understood part in both modulating and responding to the Earth's climate, and satellite remote sensing offers the best hope of studying and monitoring them. Microwave techniques are especially promising, since the ice sheets are frequently obscured by cloud, and the polar regions are subject to long periods of darkness. The main parameters that can be studied from satellite sensor observations are surface elevation and elevation change, the position and motion of ice fronts and surface features, surface velocity and strain rate, surface temperature, the extent of summer melt zones, and snow accumulation. No satellite-based technique currently permits the direct measurement of the thickness of an ice sheet, though this can be measured by surface or airborne radio echo-sounding.

Useful measurements of surface elevation require an accuracy of 5 m or better, achievable using stereophotographic methods or a *radar altimeter*. Absolute measurements of surface elevation require that the data be corrected for local topography, principally surface slope. However, changes in surface elevation can be assessed without correction for these effects since the local corrections are approximately the same provided altimeters of similar design are used. Unfortunately, intercomparison between the *GEOS-3* (1975 to 1978), *Seasat* (1978) and *Geosat* (1985 to 1990) altimeters is difficult because of their different design, but the *ERS-1* altimeter and its proposed successors should allow long-term (5 years plus) changes in ice sheet elevation to be assessed to an accuracy of ± 0.5 m or better. The use of *laser altimeters* and *interferometric SAR* will reduce these uncertainties in the future and should allow changes to be measured more reliably.

Visible-wavelength and *synthetic aperture radar* images have the potential to map ice fronts, icebergs and surface features such as crevasses, ice rises, ablation areas and glacier grounding lines to a useful accuracy of 1000 m and to measure changes in position to an accuracy of 100 m. Past coverage has been very limited, and it is still true that the far side of the Moon is better mapped than many parts of Antarctica. Surface velocities can be measured in the same way, by remapping observable features approximately once a year, though greater accuracy, suitable also for measuring surface strain rates, is promised by the use of *laser profilers* with surface *radar transponders* and by interferometric SAR.

Surface temperatures are probably best determined using *thermal infrared radiometry*, though at present the difficulty of making accurate corrections for atmospheric effects and uncertainty over seasonal variations in the *emissivity* of snow surfaces means that the accuracy is only a few K, whereas an accuracy of ±1 K or better is required.

Snow accumulation and summer melt zones can be determined using *passive microwave radiometry*, possibly in conjunction with *thermal infrared radiometry*, since the emissivity of a snow surface is strongly dependent on both grain size (which is influenced by accumulation rate and temperature) and wetness. The relationship between microwave emissivity and accumulation rate is still only empirical and poorly validated, but the signal from melting is large (typically a 60 K increase in brightness temperature) and unambiguous. It is also possible that *synthetic aperture radar* and *scatterometry* can be used to determine snow accumulation and summer melting, through the relationships between volume scattering and grain size distribution for dry snow and between surface scattering and surface roughness for wet snow. However, these techniques are also poorly validated at present.

See also *glaciers*.

IDCS (Image Dissector Camera System) U.S. visible-wavelength imaging radiometer, carried on *Nimbus-3* and *-4* satellites. <u>Waveband</u>: 0.45–0.65 μm. <u>Spatial resolution</u>: 2.2 km at nadir. <u>Swath width</u>: 1400 km.

IFOV See *spatial resolution*.

IHS display A convention for the display of a digital image, in which three channels of data (e.g. three spectral bands) are displayed as the **intensity** or brightness (I), **hue** (H) and **saturation** (S) of a colour display. The hue is a measure of the dominant wavelength of the perceived colour, and the saturation is a measure of the purity of the colour, such that the saturation decreases as more white is mixed with the colour. I and S usually take values between 0 and 1. The hue H is expressed as an angle between 0 and 360°, with 0 corresponding to red, 120° to green and 240° to blue. Several approaches to the conversion between (I, H, S) and the corresponding (R, G, B) values of an *RGB display* are possible. One such scheme is described below. If an RGB display can represent each of the three primary colours red, green and blue as integers from 0 to N (e.g. $N = 255$ for a 24-bit colour display),

$$R = NI(1 - Sr(H)),$$

$$G = NI(1 - Sg(H)),$$

$$B = NI(1 - Sb(H)).$$

The functions $r(H)$, $g(H)$ and $b(H)$ are defined as follows:

$$r(H) = t\left(2 - \frac{|H - 180|}{60}\right),$$

$$g(H) = t\left(\frac{|H - 120|}{60} - 1\right),$$

$$b(H) = t\left(\frac{|H - 240|}{60} - 1\right),$$

where $t(x)$ is a truncation function such that $t(x) = 0$ for $x < 0$, $t(x) = x$ for $0 \leq x \leq 1$, and $t(x) = 1$ for $x > 1$. For example, $(I, H, S) = (0.5, 25°, 0.75)$ gives $(R, G, B) = (128, 72, 32)$ if $N = 255$.

The reverse transformation, to obtain (I, H, S) from (R, G, B), can be performed as follows:

$$I = \frac{\max(R, G, B)}{N},$$

$$S = 1 - \frac{\min(R, G, B)}{NI}.$$

The hue angle H can then be found by calculating the values of the functions $r(H)$, $g(H)$ and $b(H)$ (unless $S = 0$, in which case it is undefined):

$$r = \frac{1}{S} - \frac{R}{NIS},$$

$$g = \frac{1}{S} - \frac{G}{NIS},$$

$$b = \frac{1}{S} - \frac{B}{NIS},$$

and distinguishing the following cases:

$$r = 0, b = 1, \qquad H = 60(1 - g),$$
$$g = 0, b = 1, \qquad H = 60(1 + r),$$
$$r = 1, g = 0, \qquad H = 60(3 - b),$$
$$r = 1, b = 0, \qquad H = 60(3 + g),$$
$$g = 1, b = 0, \qquad H = 60(5 - r),$$
$$r = 0, g = 1, \qquad H = 60(5 + b).$$

IHS values can be used in image processing in a variety of ways. In an **IHS-transformation**, the RGB values are converted to IHS values. These values are then assigned to the R, G and B channels of the display unit to produce a false-colour composite. A second possibility is to convert RGB values to IHS values, to perform suitable transformations on some or all of these (for example, *contrast enhancement* of the intensity values), and then to re-transform to RGB values for display. Thirdly, IHS values have proved particularly useful in combining images from different sources, especially when the images have different spatial resolutions. Typically the highest-resolution data are used to define the intensity value, with two other channels of data being used to generate the hue and saturation values.

IKAR-Delta (Ikarus scanning microwave radiometer) Russian scanning passive microwave radiometer, carried on Priroda module of *Mir-1* space station. Frequencies: 7.5, 22.2, 37.5, 100 GHz. Spatial resolution: 50 km at 7.5 GHz to 5 km at 100 GHz. Incidence angle: 40°. Swath width: 400 km. Sensitivity: 0.15 K at 7.5 GHz to 1.5 K at 100 GHz.

IKAR-N (Ikarus nadir microwave radiometer) Russian scanning passive microwave radiometer, carried on Priroda module of *Mir-1* space station. Frequencies: 5.0, 13.3, 22.2, 37.5, 100 GHz. Spatial resolution: 75 km at 5 GHz to 5 km at 100 GHz. Swath width: 750 km at 5 GHz to 60 km at 100 GHz.

IKAR-P (Ikarus panoramic microwave radiometer) Russian scanning passive microwave radiometer, carried on Priroda module of *Mir-1* space station. Frequencies: 5.0 and 13.3 GHz. Polarisation: H and V. Spatial resolution: 75 km. Incidence angle: 40°. Swath width: 750 km. Sensitivity: 0.15 K.

IKI Institut Kosmicheskikh Issledovanniy, the Space Research Institute of the Russian Academy of Sciences.

URL: http://arc.iki.rssi.ru/Welcome.html

Ikonos-1 US satellite, planned for launch in June 1998 and operated by Space Imaging Inc. and *EOSAT*. Objectives: High-resolution Earth observation for commercial use. Orbit: Circular *Sun-synchronous LEO* at 680 km altitude, inclination 98.2°. Principal instrument: *Carterra-1*.
Ikonos-1 was formerly known as Carterra-1.

ILAS (Improved Limb Atmospheric Spectrometer) Japanese near infrared and thermal infrared Sun-viewing *limb sounder*, for vertical profiling of atmospheric density, temperature, aerosols, H_2O, CO_2, O_3, CH_4, NO_2 and other gases, carried on *ADEOS* satellite. Wavebands: 0.753–0.784, 5.99–6.78, 6.21–11.77 μm. Spatial resolution: 13 km (horizontal), 2 km (vertical). Height range: 10–60 km.
An improved instrument, ILAS-II, is proposed for ADEOS-II: Wavebands: 0.753–0.784 μm (1024 channels with 0.1 nm resolution); 3.0–5.7 μm (22 contiguous channels); 6.21–11.76 μm (44 contiguous channels); 12.78–12.85 μm (22 contiguous channels). Spatial resolution: 13 km (horizontal); 1 km (vertical). Height range: 10–60 km.

URL: http://hdsn.eoc.nasda.go.jp/guide/guide/satellite/sendata/ilas_e.html

Image In remote sensing, a two-dimensional representation of the two-dimensional variation of intensity emitted by or reflected from the Earth's surface. In *photographic systems* the spatial variation is recorded in essentially analogue form, but in all other imaging systems it is quantised into *pixels*. Each pixel may record just a single measure of the intensity originating from the corresponding region of the Earth's surface, or several measurements (such as the intensity in several different wavebands or with different polarisation states).

Image Dissector Camera System See *IDCS*.

Image format The sequential arrangement of the data representing a digital *image* in a storage medium such as a magnetic tape or compact disc. The main formats in use are **band interleaved by pixel (BIP)**, **band interleaved by line**, and **band sequential (BSQ)**. *Run-length coding* is also sometimes used.

In both BIP and BIL formats, the data are stored a line (row) at a time. In the BIP format, the data within a line are stored in the following sequence:

pixel 1 band 1; pixel 1 band 2; ... pixel 1 band n;
pixel 2 band 1; pixel 2 band 2; ... pixel 2 band n;
...
pixel m band 1; pixel m band 2; ... pixel m band n,

whereas in BIL format the sequence is

pixel 1 band 1; pixel 2 band 1; ... pixel m band 1;
pixel 1 band 2; pixel 2 band 2; ... pixel m band 2;
...
pixel 1 band n; pixel 2 band n; ... pixel m band n.

In BSQ format, each band of the data is stored as a separate file. Within the file, the sequence is

pixel 1 line 1; pixel 2 line 1; ... pixel m line 1;
pixel 1 line 2; pixel 2 line 2; ... pixel m line 2;
...
pixel 1 line l; pixel 2 line l; ... pixel m line l.

In all of these examples, the image is assumed to consist of l lines (rows) each consisting of m pixels. Each pixel is specified by n digital numbers (for example, corresponding to n spectral bands).

Image processing Manipulation of the data contained in an *image*, normally by means of a digital computer, to extract quantitative information or to emphasise features of interest. Image processing is normally considered to include pre-processing, contrast enhancement, image transformation, spatial filtering, and classification, though it can also include the combination of more than one image.

The division between pre-processing and processing is somewhat artificial, but it is conventional to consider *geometric correction* and *radiometric correction* (including *atmospheric correction* and *destriping*) as the main pre-processing operations, resulting in an image that corresponds geometrically to a chosen coordinate system or map projection, and radiometrically to appropriate physical units such as surface reflectance or radiance.

Contrast enhancement operations optimise the use of the computer's display unit to represent the image data, generally without modifying the data themselves. Similarly, *density slicing* and pseudocolour transformations optimise the use of a colour display to represent a single-band image. Image transformation, on the other hand, modifies the image data.

Simple image transformations include arithmetic operations (addition, multiplication etc.) on the different bands of an image, or on different images that have been co-registered. More sophisticated image transformations include the *principal components* and *canonical components* transformations, and the derivation of *vegetation indices*. These are all pixel-based operations, i.e. each *pixel* is processed without reference to the other pixels in the image. Spatial image transformations, in which the spatial frequency content of the data is the parameter of interest, include the *Fourier transform* and the *Hadamard transform*.

Spatial filtering operations change the spatial frequency content of the data, for purposes such as the suppression of random noise, smoothing, sharpening, or detecting edges in the image. These operations can either be performed directly on the image, by convolving it with a suitable filter function (see *convolution filtering*), or as a frequency-domain filter by calculating the Fourier or Hadamard transform of the image, performing a suitable filtering operation, and then re-transforming.

The final step in image processing is usually that of *classification*, in which each pixel of the image is assigned to one of a finite number of classes (for example, different types of land cover) on the basis of suitable decision rules.

Imaging radar See *radar*.

Imaging radiometer A type of *electro-optical sensor*, operating in the optical or infrared part of the electromagnetic spectrum and providing radiance data for a small number of spectral bands for each *pixel* of the image. If the number of spectral bands is larger, the instrument can instead be termed a *multispectral imager*.

Many of the earlier spaceborne sensors had only one or two spectral bands, as a result of the technical difficulty of providing more. However, imaging radiometers continue to be used, often to meet some more specialised requirement than for the multispectral imagers. As with the multispectral imagers, a wide range of spatial resolutions is represented by this class of instrument.

High spatial resolution (5 m) is provided by the *PAN* instrument on *IRS-1C*, with a swath width of 71 km, and would have been available from the *HSI* instrument on *Lewis*. Even higher spatial resolution (1–3 m or finer) will be provided by the narrow-swath *Carterra-1*, *Earthwatch Imager*, *Orbview-3* and *QuickBird*. Examples of intermediate resolution (50 to 500 m) instruments are *MSU-S*,

WiFS and *WFI* (swath widths typically 800 to 1500 km). Low-resolution (>500 m) instruments include *HCMR* (thermal inertia measurements), *HRIR* (night-time temperature measurements), *Klimat*, *OLS* (day and night cloud cover), *LIS* (lightning), *THIR*, *SSC* (discrimination of snow from cloud) and *VHRR*. These instruments have swath widths generally in the range 700 to 3000 km.

Imaging system Remote sensing instrument that generates an *image*. Compare *non-imaging system*.

IMG (Interferometric Monitor for Greenhouse Gases) Japanese infrared nadir-looking spectrometer on *ADEOS* satellite. Waveband: 3.3 to 14.0 µm. Spectral resolution: 3 GHz (1 nm at 10 µm) Fourier transform (Michelson interferometer) spectrometer. Horizontal resolution: 8 km IFOV, but degraded by the 10 s scan time. Vertical resolution: 2 to 6 km depending on the profiled species. IMG provides profiles of atmospheric CO_2, CH_4 and N_2O.

Impedance The impedance Z of a medium with respect to electromagnetic radiation is given by

$$Z = \sqrt{\frac{\mu_r \mu_0}{\varepsilon_r \varepsilon_0}} = Z_0 \sqrt{\frac{\mu_r}{\varepsilon_r}},$$

where μ_r is the relative magnetic permeability and ε_r is the relative electric permittivity (dielectric constant) of the medium. μ_0 is the permeability of free space, ε_0 the permittivity of free space, and Z_0 the impedance of *free space*.

Improved Atmospheric Sounding Interferometer See *IASI*.

Improved Limb Atmospheric Spectrometer See *ILAS*.

Improved Stratospheric and Mesospheric Sounder See *ISAMS*.

Improved TIROS Operational Satellite See *ITOS*.

Impulse, specific See *orbital manoeuvres*.

Incidence angle The angle between the propagation direction of radiation emitted by an *active system* and the normal to the surface on which it is incident. The incidence angle is zero for normal incidence, 90° for grazing incidence.

Inclination The inclination of a satellite *orbit* is the angle between the orbital plane and the plane of the Earth's equator. The inclination is defined to be less than 90° if the satellite moves in its orbit in a *prograde* direction, and greater than 90° if its motion is *retrograde*. For an inclination of 90°, the orbit passes directly over the Earth's poles.

Indian Remote Sensing Satellite See *IRS-1*.

Information content From the point of view of information theory, the information content of a set of N numbers drawn from a set of discrete values (e.g. integers) can be derived from the frequency histogram of the values. If each number can take one of M possible values (e.g. $M = 2^n$ for n-bit data), and value i occurs n_i times within the set of N numbers, the theoretical information content of the whole set is given by

$$N \log_2 N - \sum_{i=1}^{M} n_i \log_2 n_i \quad \text{bits,}$$

where

$$N = \sum_{i=1}^{M} n_i.$$

Data compression techniques (e.g. the *Huffman code*) can code the data in such a way that the number of bits needed to specify it is only slightly greater than this theoretical value.

Infrared A region of the *electromagnetic spectrum*, corresponding to wavelengths between about 0.78 μm and 1 mm. In remote sensing applications, the most useful part of the infrared spectrum lies between 0.78 μm and about 15 μm. The thermal infrared region is often subdivided into the **near infrared** (0.78 μm to about 1.2 μm), **middle infrared** (about 1.2 μm to 3 μm) and **thermal infrared** (3 μm to 15 μm). Near infrared radiation, sometimes called **photographic infrared**, can be detected on suitable photographic film, whereas middle and thermal infrared radiation is detected using *electro-optical systems*. Solar radiation contains substantial amounts of near infrared and middle infrared radiation (sometimes collectively referred to as **reflective infrared**), and many remote sensing systems are designed to respond to reflected solar radiation in these bands, especially for monitoring *vegetation* and surface water. Thermal infrared radiation is emitted by all bodies in quantities dependent on their temperature and *emissivity*, and detection of this radiation is mainly used for determining surface and atmospheric temperatures (see *thermal infrared radiometry, atmospheric sounding*).

An alternative nomenclature for infrared radiation, used in this book, defines near infrared to be between 0.78 and 3 μm (i.e. the reflective infrared region); thermal infrared from 3 μm to about 30 μm; and **far infrared** from 30 μm to 1 μm.

Infrared Interferometer Spectrometer See *IRIS*.

Infrared Multispectral Scanner See *IR-MSS*.

Infrared radiometers Carried on the Spektr module of *Mir-1:*

Instrument	Waveband (µm)	Spatial resolution (km)	Spectral resolution
Jausa	1.8–3.0	5	1.2 µm
Neva-3	1.8–3.0	1	1.2 µm
Phoenix	2.61–2.63	3 × 9	12 GHz
Neva-5	3.0–5.0	1 × 2	2.0 µm
Volkhov-1	5–22	60	300 GHz
Volkhov-2	5–22	2	480 GHz

Infrared Sensor See *IR*.

Infrared Temperature Profiling Radiometer See *ITPR*.

INPE Instituto de Pesquisias Espaciais, the Brazilian Space Agency.
Address: São José dos Campos, Avenida dos Astronautas 1758, São José dos Campos, SP, 12227-010, Brazil.

URL: http://www.inpe.br/

Insat (Indian National Satellite) Indian satellite series, operated by the Departments of Space, Telecommunications and Meteorology, as well as All-India Radio (for TV transmissions). Insat-1A launched April 1982, 1B August 1983, 1C July 1988 (failed November 1989), 1D June 1990, 2A July 1992, 2B July 1993. Objectives: Operational meteorology, collection and relay of weather data, TV relay. Orbit: *Geostationary* (longitude 74° E). Principal instruments: *VHRR*. Insat satellites also carry data collection packages, similar in operation to *Argos*. Insat-2 satellites also carry a search and rescue system.

URL: http://www.fas.org/spp/guide/india/earth/insat2_eo.htm

Instantaneous field of view See *spatial resolution*.

Integral equation model A surface *scattering model* for randomly rough surfaces, valid at all frequencies. It is an expression relating the scattering coefficient to frequency, polarisation, incident and scattered angles, surface dielectric constant, surface correlation function and RMS surface height variation. The surface current density responsible for scattering is derived from the integral equations governing it. These are solved approximately in two steps.

First, the integral equations are recast in a form such that the surface current density is expressed as a sum of two terms. The first term is the Kirchhoff current density and the second term is an integral involving the unknown surface current density.

Second, the unknown surface current density under the integral sign in the second term is estimated using the Kirchhoff surface current density. As a result, a completely known surface current is obtained and hence the reradiated field and finally the scattering coefficient can be computed in terms of it.

A major feature of this model is that it is not restricted to a frequency range. Unlike the *small perturbation model* and the *Kirchhoff model*, it can predict the behaviour of the scattering coefficient as a function of frequency, and is sensitive to the imaginary part of the surface dielectric constant. Thus, this model is useful for the retrieval of surface parameters. Under low and high frequency conditions, the predictions of this model are in agreement with, respectively, the small perturbation model and the Kirchhoff model.

Intensity See *radiant intensity, IHS display*.

Interferometric Monitor for Greenhouse Gases See *IMG*.

Interferometric SAR The combination of two or more *synthetic aperture radar* (SAR) images of the same area to obtain accurate topographic or motion information about the target area. This information is contained in the phase differences between the images, so the *coherence* of the images must be preserved in order to make a phase-difference image (interferogram). Interferometric SAR can be performed using repeated passes (orbits) of the same sensor, or an array of two or more SAR sensors. The distance between the positions of the SAR sensor used to generate the two images from which the interferogram is calculated is termed the **baseline**. This must not exceed some maximum value (typically of the order of 1 km for the *ERS* instruments) set by the SAR's viewing geometry, frequency and spatial resolution, otherwise the phase difference will change too rapidly from one pixel to the next. The successful generation of an interferogram also requires that the physical properties of the target area do not change between the times when the images are obtained, which is more difficult to ensure in the case of repeat-orbit interferometry.

Interferometric SAR can be used to determine surface topography and bulk translation of parts of the surface between images. For topographic measurements the baseline B should be as large as possible, since the interference fringes are separated in altitude by a distance that is inversely proportional to B. For example, for the *ERS* instruments the elevation contour interval per fringe is roughly $8000/B$ giving an 80-m contour interval for a typical baseline of 100 m. On the other hand, surface motion produces effects that are almost independent of the baseline, so short baselines are preferred in order to minimise the topographic contribution to the fringes. For two images acquired from the same location but at different times, a displacement of half a wavelength in the position of the surface will produce an interference fringe. Accuracies of better than one wavelength are thus possible. For both topographic and motion measurements, higher accuracies (measurement of fractions of a fringe) can be achieved if the signal-to-noise ratio is sufficiently large.

The major disadvantages of interferometric SAR are the technical complexity of the image processing, and the stringency of the conditions that must be satisfied if a useful interferogram is to be generated.

Intermediate Thermal Infrared Radiometer See *ASTER*.

Ionosphere Region of the Earth's upper *atmosphere*, extending from a height of approximately 50 km to over 1000 km, in which ionisation of nitrogen and oxygen atoms by solar ultraviolet and X-ray radiation results in high concentrations of ions and free electrons. The ionisation is significantly greater on the sunlit side of the Earth than on the night-side; it also increases as a result of increased solar activity. The figure illustrates typical day and night-time variations of free-electron density with height. It also shows the approximate locations of the D, E, F1 and F2 layers into which the ionosphere is conventionally divided.

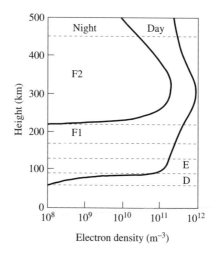

Electron density (m^{-3})

The ionosphere has a negligible effect on the propagation of visible-wavelength and infrared radiation. At radio frequencies it introduces propagation delays which can be significant for high-precision *radar altimeter* measurements and for satellite navigation systems such as *GPS*. At microwave frequencies, the propagation delay for a pulse travelling vertically through the ionosphere, compared to the propagation time for a pulse travelling the same distance at the speed of light, is

$$\frac{e^2}{2\pi m f^2 \varepsilon_0 c} \int_0^\infty N \, dz,$$

where e is the electronic charge, m is the electron mass, f is the frequency, ε_0 is the permittivity of free space, c is the speed of light, and N is the electron density (electrons per unit volume) at height z. The quantity

$$\int_0^\infty N \, dz$$

is called the **total electron content** (TEC), and has a typical daytime value of 3×10^{17} m^{-2} and a typical night-time value about 10 times less. The pulse delay can be written in terms of the frequency and the TEC as

$$13.4 \left(\frac{f}{\text{GHz}}\right)^{-2} \left(\frac{\text{TEC}}{10^{17}\,\text{m}^{-2}}\right) \text{ns.}$$

Absorption of microwave radiation by the ionosphere is negligible.

IR

1. Usual abbreviation for infrared (see *electromagnetic spectrum*).
2. (Infrared sensor) Russian thermal infrared imaging radiometer, carried on *Meteor-1* and *-2*, and *Meteor-Priroda*, satellites. Waveband: 8–12 μm. Spatial resolution: 15 km at nadir. Swath width: 1000 km.

IRIS (Infrared Interferometer Spectrometer) U.S. nadir-looking infrared spectrometer, for atmospheric profiling, carried on *Nimbus-3* and *-4*. Wavebands: 5–20 μm (Nimbus-3), 6.3–25 μm (Nimbus-4). Spatial resolution: 140 km.

IR-MSS (Infrared multispectral scanner) Chinese optical/near infrared/thermal infrared imaging radiometer, to be carried on *CBERS* satellite. Wavebands: 0.5–1.1, 1.55–1.75, 2.08–2.35, 10.4–12.5 μm. Spatial resolution: 78 m (156 m for thermal infrared band). Swath width: 120 km.

Irradiance The total electromagnetic power incident from all directions on a plane surface area dA is given by

$$E\, dA$$

where E is the irradiance. The SI unit of irradiance is $\text{W}\,\text{m}^{-2}$. If the incoming *radiance* on the surface is $L(\theta, \phi)$ in the direction which makes an angle θ with the surface normal and has azimuth angle ϕ, the irradiance is given by

$$E = \int_{\theta=0}^{\pi/2} \int_{\phi=0}^{2\pi} L(\theta, \phi) \sin\theta\, d\theta\, d\phi.$$

See also *radiant exitance*.

IRS-1 (Indian Remote Sensing Satellite) Indian satellite series operated by *ISRO*. The programme began with the launch of IRS-1A in March 1988. The satellites have a nominal lifetime of 3 years. Objectives: Operational environmental monitoring. Orbit: Circular *Sun-synchronous LEO* at 904 km altitude (817 km for IRS-1C and -1D). Period 103.2 minutes (101.3 minutes for IRS-1C and -1D); inclination 99.1° (98.7° for IRS-1C and -1D); equator crossing time 10:26. *Exactly-repeating orbit* (307 orbits in 22 days [IRS-1A, -1B], 341 orbits in 24 days [IRS-1C, -1D]). Principal instruments: *LISS-1* (IRS-1A, -1B), *LISS-2* (IRS-1A, -1B), *LISS-3* (IRS-1C, -1D), *PAN* (IRS-1C, -1D), *WiFS* (IRS-1C, -1D).

IRS-1E failed on launch in September 1993. IRS-1C was launched in December 1995, and IRS-1D in September 1997.

URL: http://www.spaceimage.com/home/products/carterra/irs/constell.html

IRS-P (Indian Remote Sensing Satellite) Indian–German satellite series, jointly operated by *ISRO* and *DLR*, begun with the launch of IRS-P2 in October 1994. IRS-P4 was launched in 1996 and -P5 in 1997. Further satellites in the series are planned. Objectives: Operational environmental monitoring. Orbit: Circular *Sun-synchronous LEO* at 817 km altitude. Period 101.3 minutes; inclination 98.7°; equator crossing time 10:30 (descending node). *Exactly-repeating orbit* (341 orbits in 24 days). Principal instruments: *LISS-2* (IRS-P2), *LISS-3* (IRS-P4), *MOS-A* (IRS-P3), *MOS-B* (IRS-P3), *MOS-C* (IRS-P3).

URL: http://www.spaceimage.com/home/products/carterra/irs/constell.html

ISAMS (Improved Stratospheric and Mesospheric Sounder) UK nadir-viewing infrared filter radiometer carried by *UARS* satellite. Waveband: 4.6–16.6 μm.

ISAMS was an improved version of *SAMS*, used to measure profiles of ozone, water vapour, methane, CO_2 and nitrogen compounds in the upper atmosphere. It ceased operation in July 1992.

URL: http://www.atm.ox.ac.uk/group/isams/

Isodata algorithm An iterative *clustering* algorithm, also called the K-means or migrating means technique. The user specifies the number of clusters to be found. The algorithm assigns nominal centre coordinates in feature space to each of these clusters, either at random or using any available knowledge about the clusters. Each pixel in the image is assigned to the cluster to the centre of which it is closest in feature space. The cluster centre coordinates are then recalculated from the means of the feature vectors of the pixels that have been assigned to them, and the process is repeated. It is normally terminated when the number of pixels reassigned to different clusters falls below some threshold.

ISRO Indian Space Research Organisation, the Indian Space Agency.

ISSA (International Space Station Alpha) Satellite planned for launch into *LEO* in 2001, with an inclination of about 52°, as part of the *EOS* programme. ISSA will carry the *SAGE III* instrument, a mission formerly planned for one of the EOS-Aero satellites.

URL: http://issa-www.jsc.nasa.gov/

Istok-1 Russian infrared spectrometer carried on *Mir-1*. Operates in nadir or limb-sounding modes. Wavebands: 64 channels between 3.6 and 16 μm. Spectral resolution: 0.13 μm (below 8 μm), 0.25 μm (above 8 μm). Spatial resolution: 0.8 × 6 km.

The instrument is used for atmospheric profiling of temperature, pressure, water vapour and other constituents including ozone, and for measuring *SST*.

ITIR See *ASTER*.

ITOS (Improved TIROS Operational Satellite) U.S. satellite, operated by *NOAA*, launched January 1970. Objectives: Operational meteorological observations (successor to *TIROS*). Orbit: Circular *LEO* at 1650 km altitude. Period 115 minutes; inclination 102°. Principal instruments: *APT, AVCS, FPR, SR*.

ITPR (Infrared Temperature Profiling Radiometer) U.S. nadir-viewing infrared radiometer, for atmospheric profiling, carried by *Nimbus-5*. Wavebands: seven channels between 3.7 and 20 µm. Spatial resolution: 32 km.

ITS-7D Russian nadir-viewing infrared spectrometer, for atmospheric profiling, carried on *Mir-1*. Waveband: 4.0 to 8.0 µm. Spectral resolution: 0.15 µm. Spatial resolution: 60 × 240 km.

Japanese Earth Resources Satellite See *JERS-1*.

Jason-1 US/French satellites, operated by *NOAA*, *NASA* and *CNES*, planned for launch in 2000 as part of the *EOS* programme. Nominal lifetime: 5 years. Objectives: Dedicated altimetry missions for ocean topography, geoid measurement etc. Orbit: As for *Topex-Poseidon*: Circular *LEO* at 1334 km altitude. Period 112 minutes; inclination 66°. Principal instruments: *DORIS, DFA, AMR*.

Jason-1 was formerly known as TPFO (*Topex-Poseidon* Follow-On) and as **EOS-Alt**, the *EOS* radar altimeter satellite.

URL: http://topex-www.jpl.nasa.gov/jason/jason.html

Jausa See *Infrared radiometers*.

Jeffries–Matusita distance See *separability*.

JERS-1 (Japanese Earth Resources Satellite) Japanese satellite, operated by *NASDA* and MITI, launched February 1992. Objectives: Experimental and operational environmental mapping and monitoring. Orbit: Circular *Sun-synchronous LEO* at 568 km altitude. Period 96.1 minutes; inclination 97.7°. *Exactly-repeating orbit* (659 orbits in 44 days). Principal instruments: *OPS, SAR*.

JERS-1 is also known as Fuyo.

URL: http://hdsn.eoc.nasda.go.jp/guide/guide/satellite/satdata/jers_e.html

JM distance See *separability*.

JPOP See *ADEOS II*.

K$_a$ band Subdivision of the microwave region of the *electromagnetic spectrum*, covering the frequency range 33 to 36 GHz (wavelengths 8.3 to 9.1 mm).

KAP-350 *Photographic system* carried on *Mir-1*. Focal length: 350 mm. Image format: 180 × 180 mm. Coverage: 180 km × 180 km (from 350 km altitude). Spatial resolution: 30 m (from 350 km altitude).

Kappa value See *error matrix*.

Karhunen–Loève transformation See *principal components*.

KATE-140 Visible-wavelength camera system carried by *Salyut* missions and by Priroda module of *Mir* space station. Operating wavelengths 0.5 to 0.7 µm. Resolution (from 400 km altitude) 60 m; swath width (from 400 km altitude) 270 km.

Kauth–Thomas transform See *tasselled-cap transformation*.

K band Subdivision of the microwave region of the *electromagnetic spectrum*, covering the frequency range 10.9 to 36 GHz (wavelengths 8 to 28 mm).

Kernel See *convolution operator*.

Keyhole See *Argon, Corona, Lanyard*.

KFA-200 *Photographic system* carried on *Resurs-F*. Focal length: 200 m. Wavebands: 0.51–0.60, 0.60–0.70, 0.70–0.84 µm (filters). Format: 180 × 180 mm. Coverage: 245 km × 245 km (from 275 km altitude). Spatial resolution: 30 m (from 275 km altitude).

KFA-1000 *Photographic system* carried on *Mir-1* (Kristall module) and *Resurs-F*. Focal length: 1000 mm. Waveband: 0.57–0.80 µm (filter). Image format: 300 × 300 mm. Coverage: 105 km × 105 km (from 350 km altitude). Spatial resolution: 6–20 m (from 350 km altitude).

KFA-1000 is primarily intended for reconnaissance rather than mapping, since the images are subject to some radial distortion.

URL: http://www.eurimage.it/Products/KFA_1000.html

Kirchhoff law Kirchhoff's law for radiation states that, at a given temperature and wavelength, the *emissivity* of a body is equal to its absorptance, defined as the ratio of absorbed power to incident power.

Kirchhoff model A surface *scattering model* for randomly rough surfaces, also called the physical optics model. It uses the tangent plane approximation (physical optics approximation) to determine an approximate expression for the surface current density which is responsible for scattering. This approximation assumes that the surface field at any point on the surface can be computed with respect to an infinite plane tangent to the point instead of the real surface. This assumption is valid if the plane area around every local point is at least one wavelength in size or, equivalently, the radius of curvature at every point is larger than a wavelength. For this reason, a suitable surface for this model has been referred to as a gently undulating surface.

With the surface current density known, the scattered field is then computed in terms of it, and finally the scattering coefficient is calculated from the scattered field. The dielectric property of the surface appears only through the magnitude squared of the *Fresnel reflection coefficient*, and hence the model cannot distinguish the real from the imaginary part of the dielectric constant. In the high-frequency limit, the model loses its dependence on frequency, RMS height variation and surface correlation length, and shows that scattering is proportional to the probability density function of the surface slopes in orthogonal directions.

Two variants of the Kirchhoff model are in common use: the stationary phase approximation (or geometric optics model), which is valid for rougher surfaces, and the scalar approximation. The conditions for validity of each of these are as follows:

Stationary phase	Scalar
$2k\sigma\cos\theta > \sqrt{10}$	$m < 0.25$
$kL > 6$	$kL > 6$
$L^2 > 2.76\sigma\lambda$	$L^2 > 2.76\sigma\lambda$

k is the wavenumber of the incident radiation ($=2\pi/\lambda$), σ is the RMS surface height variation, L is the correlation length of the surface height variations, θ is the incidence angle, and m is the RMS surface slope.

The (incoherent) stationary phase model is given by

$$\sigma_{HH}^0(\theta) = \sigma_{VV}^0(\theta) = \frac{|R(0)|^2 \exp\left(-\dfrac{\tan^2\theta}{2m^2}\right)}{2m^2\cos^4\theta},$$

where $R(0)$ is the Fresnel coefficient for amplitude reflection at normal incidence. The *cross-polarised* terms are zero.

The scalar approximation model is given by

$$\sigma_{pp}^0(\theta) = 2k^2\cos^2(\theta)|R_p(\theta)|^2\exp(-4k^2\sigma^2\cos^2\theta)$$

$$\times \sum_{n=1}^{\infty} \frac{(4k^2\sigma^2\cos^2\theta)^n}{n!} \int_0^\infty \rho^n(\xi)J_0(2k\xi\sin\theta)\xi\,d\xi,$$

where σ_{pp}^0 is the *backscattering coefficient* for *pp*-polarised radiation (i.e. $p = H$ or V), $R_p(\theta)$ is the Fresnel coefficient for amplitude reflection of *p*-polarised radiation at incidence angle θ, and $\rho(\xi)$ is the surface autocorrelation function. If the autocorrelation function is Gaussian with a correlation length L ($\rho = \exp[-\xi^2/L^2]$), the formula becomes

$$\sigma^0 = k^2 L^2 \cos^2 \theta |R_p(\theta)|^2 \exp(-4k^2\sigma^2 \cos^2 \theta)$$

$$\times \sum_{n=1}^{\infty} \frac{(4k^2\sigma^2 \cos^2 \theta)^n}{n!\, n} \exp(-k^2 L^2 \sin^2 \theta/n).$$

σ, L and m are then related by

$$m = \sqrt{2}\,\frac{\sigma}{L}.$$

See *integral equation model, small perturbation model*.

K$_I$ band Subdivision of the microwave region of the *electromagnetic spectrum*, covering the frequency range 15.35 to 24.5 MHz (wavelengths 12.2 to 17.4 mm).

KL-103W Russian colour *vidicon* imaging radiometer carried on Kvant-2 module of the *Mir-1* space station. Waveband: 0.4–0.8 μm. Spatial resolution: 250 m. Image size: 100 km × 125 km.

Klimat Russian thermal infrared mechanically scanned imaging radiometer, carried on *Meteor-3* satellites. Waveband: 10.5–12.5 μm. Spatial resolution: 3 km at nadir. Swath width: 3100 km. Temperature sensitivity: 0.2 K.
Klimat-2 will also have a channel at 0.65–1.0 μm and will have a spatial resolution of 1 km at nadir.

Kosmos Russian satellite series. Orbit: Circular *LEO* at 190 to 270 km altitude; inclination 71°. Principal instruments: *KVR-1000, TK-350*.

Kosmos 1870 Russian satellite, launched in July 1987 and operating for 2 years. Objective: Experimental radar remote sensing satellite as a prototype for the *Almaz* missions. Orbit: Circular at 275 km altitude. Period 92 minutes; inclination 73°. Principal instruments: *SAR*.

Kristall See *Mir-1*.

K$_u$ band Subdivision of the microwave region of the *electromagnetic spectrum*, covering the frequency range 15.35 to 17.25 GHz (wavelengths 17.4 to 19.5 mm).

Kvant-2 See *Mir-1*.

KVR-1000 Russian photographic system on *Kosmos* satellites. Spatial resolution: 2 m. Image size: 40 × 40 km.

URL: http://www.eurimage.it/Products/KVR_1000.html

LAC See *AVHRR*.

Lageos (Laser Geodynamics Satellite) Series of satellites carrying 426 laser retrore-flectors, used for ground-based laser ranging to determine the *geoid* and for studying tectonic motion. Lageos-1 was launched May 1976, Lageos-2 in October 1992. Design lifetime 50 years. Orbit: semimajor axis 12 000 km, inclination 110° (Lageos-1), 52° (Lageos-2), period 228 minutes.

LAI See *leaf area index*.

Lambert–Bouguer law (Lambert's law; Bouguer's law) If reflection and scattering effects are small compared to absorption effects, the absorption of a parallel beam of radiation passing through an absorbing medium can be described by the *absorption coefficient* γ_a, such that the intensity I of the radiation varies with distance x in the propagation direction according to

$$\frac{dI}{dx} = -\gamma_a I.$$

If γ_a varies with x, the change in intensity on passing through a slab of thickness x_0 is given by

$$I = I_0 \exp\left(-\int_0^{x_0} \gamma_a \, dx\right) = I_0 \exp(-\tau),$$

where τ is the *optical thickness* of the slab.

Attenuation obeys a similar law, and an optical thickness can also be defined for attenuation. Note that absorption and attenuation coefficients are often specified in *decibels* per unit length. The decibel value of the absorption coefficient is given by $4.343\gamma_a$, and similarly for the attenuation coefficient.

See also *refractive index, radiative transfer equation*.

Lambert equal-area projection See *map projection*.

Lambertian reflection An ideally rough surface scatters radiation according to Lambert's law, according to which the reflected *radiance* is isotropic for any illumination that is uniform across the surface. Such a surface has a constant value of the *bidirectional reflectance distribution function*.

Landsat-1 to -3 U.S. satellite series operated by *NASA*. The first three satellites in the programme had the following launch and termination dates: Landsat-1, July 1972 to January 1978; Landsat-2, January 1975 to July 1983; Landsat-3, March 1978 to March 1983. Objectives: Experimental and operational environmental mapping and monitoring, particularly of the land surface. Orbit: Circular *Sun-synchronous LEO* at 907 km altitude; period 103.3 minutes; inclination 99.1°; equator crossing time (descending node) 08:50 (Landsat-1), 09:08 (Landsat-2), 09:31 (Landsat-3). *Exactly-repeating orbit* (251 orbits in 18 days). Principal instruments: *MSS*, *RBV*. The satellites also carried data collection packages.

Landsat-1 was originally called ERTS (Earth Resources Technology Satellite).

URL: http://geo.arc.nasa.gov/sge/landsat/landsat.html

Landsat-4 to -5 U.S. satellite series, operated by *NASA* (to 1984) and *Eosat* (from 1984). Landsat-4 was launched in July 1982, Landsat-5 in March 1984. Objectives: Operational environmental mapping and monitoring. Orbit: Circular *Sun-synchronous LEO* at 705 km altitude; period 99.0 minutes; inclination 98.2°; equator crossing time 09:39 (descending node). *Exactly-repeating orbit* (233 orbits in 16 days). Principal instruments: *MSS, TM*.
Landsat-4 was placed on standby in 1993. Landsat-5 is still operational.

URL: http://geo.arc.nasa.gov/sge/landsat/landsat.html

Landsat-6 to -7 U.S. satellites, continuing the programme represented by Landsat 1–5. Landsat-6 was launched in October 1993 but failed to achieve orbit. Landsat 7 is scheduled for launch in February 1999, and will be integrated into the *EOS* programme. Orbit: Circular *Sun-synchronous LEO* at 705 km altitude; period 99.0 minutes; inclination 98.2°; equator-crossing time 10:00 (descending node). *Exactly-repeating orbit* (233 orbits in 16 days). Principal instruments: *ETM+, HRMSI*.

URL: http://geo.arc.nasa.gov/sge/landsat/landsat.html

Landsat Advanced Technology Instrument See *ETM*.

Land surface temperature See *temperature, land surface*.

Lanyard U.S. military reconnaissance satellite, operating between March and July 1963. The satellite carried a panchromatic camera (Keyhole) giving a resolution of 2 m. The data are now declassified.

URL: http://edcwww.cr.usgs.gov/dclass/dclass.html

Large Format Camera See *LFC*.

Laser Geodynamics Satellite See *Lageos*.

Laser profiler/altimeter See *lidar*.

Laser retroreflector Passive device consisting of three planar and mutually perpendicular mirrors forming the inside corner of a cube. Laser retroreflectors are mounted on some satellites to allow accurate ranging to them from laser ranging stations on the Earth's surface.

LATI See *ETM*.

Layover A type of geometric distortion inherent in *side-looking radar* and *synthetic aperture radar* imagery as a consequence of the oblique *incidence angle*.

L band Subdivision of the microwave region of the *electromagnetic spectrum*, covering the frequency range 390 MHz to 1.55 GHz (wavelengths 193 to 769 mm).

Leaf The spectral reflectance properties of a leaf in the optical and near infrared parts of the spectrum depend on its structure, its water content, and the presence of photosynthetically active pigments. In the optical part of the spectrum the effects of pigments are dominant. The primary leaf pigments are *chlorophyll*-a and -b, carotene and xanthophyll. The chlorophylls absorb blue and red light, giving leaves their characteristic green colour; carotene and xanthophyll absorb blue and green light.

At near infrared wavelengths, the reflectance properties of leaves are dominated by their physiological structure. Multiple scattering of radiation by the interfaces between cells and air spaces in the mesophyll, and generally low absorption, causes the characteristically high reflectance observed between about 0.65 μm and 1.2 μm. At longer wavelengths, increased absorption by water (mainly due to the absorption bands at 1.4, 1.9 and 2.7 μm) reduces the reflectance.

See also *vegetation*.

Leaf area index The leaf area index (LAI) of a vegetated area is defined as the total *leaf* area per unit area of ground surface. *Vegetation indices* generally show strong correlation with the LAI, particularly when the vegetation canopy is dense so that the reflectance properties of the canopy are dominated by those of the leaves.

Legal and international aspects Since the launching of the first dedicated remote sensing satellite (*Landsat-1*) by the U.S.A. in 1972, many other countries and organisations have become involved in the acquisition of data about the Earth and its environment from space. These data have been used to monitor environmental problems and to provide input to resource assessment, as well as in global and regional surveillance activities.

The Outer Space Treaty, adopted by the United Nations General Assembly in 1966, affirmed the use of the outer space environment by and in the interests of all mankind. Member states of the UN, by adhering to the articles of the

Treaty, accept that there are no national boundaries in outer space. This, in turn, has encouraged and promoted international cooperation in space exploration, including remote sensing, and has created the conditions for the peaceful remote appraisal of natural resources. These include the remote acquisition of information on a nation's territory by governmental or private entities that have the necessary technical capability. It is thus understandable that many countries, particularly those that are not 'space capable', have been and are still concerned that these Earth Observation systems will be dominated by the private sector and a few governments, with attendant commercial, political, legal and security implications.

These concerns have been heightened by the end of the cold war, and the declassification of the high spatial resolution remote sensing systems that were formerly the prime tools for military surveillance. In 1986, when the Remote Sensing Principles were enunciated, the highest spatial resolution available from non-military satellites was 80 m. Today, spatial resolutions of finer than 10 m are available. The information acquired by such systems may have security implications, particularly when it may be of strategic value and is commercially available to third parties without the concurrence of the sensed state.

Notwithstanding such concerns, it is now universally accepted that the Earth is a unified system in which events in one location have repercussions in other parts of the world. A global effort to acquire information through an international remote sensing system that could be used to resolve issues such as environmental pollution, deforestation, food security and natural disasters, is a logical step. The 1992 UNCED conference held in Rio endorsed the establishment of a global Earth Observing system, and regional and international collaborations such as the *ERS* and *EOS* programmes represent moves in this direction. Furthermore, international cooperation in the acquisition and use of environmental and natural resource information has become inevitable, partly because an increasing number of countries are becoming 'space capable', and partly because of the high cost of undertaking space business.

In the forefront of a number of efforts to achieve global cooperation in remote sensing is *CEOS* (the Committee on Earth Observation Satellites), established in 1984 to coordinate space programmes dedicated to monitoring the global environment. Specifically, CEOS aims at optimising the benefits of Earth Observation from space through the cooperation of its members in mission planning, particularly with respect to the issues of data availability, compatibility, complementarity, production formats, services and applications. Current efforts aimed at addressing global environmental concerns are rooted in the United Nations General Assembly Resolution A/RES/1721(XVI) of 20 December 1961, which called for a study of 'measures to advance the state of atmospheric sciences and technology in order to improve weather forecasting capabilities and to further the study of the basic physical processes that affect climate'. Major international programmes, with substantial remote sensing elements, responding to this resolution, include GARP (the Global Atmospheric Research Programme) and IGBP (the International Geosphere-Biosphere Programme).

Steps have also been taken towards the goals of international cooperation, availability of participatory opportunities, and assistance in remote sensing. For example, *ESA* and Canada are assisting a number of non space-capable countries to gain an understanding of the use of remote sensing data in the management of the global environment as well as in the assessment of their own resources. In March 1994, the United States issued a Cooperative Agreement Notice soliciting proposals for public use, via the Internet, of the very large remote sensing databases in the U.S.A. One major concern is that remote sensing activities may not be economically viable or driven by the needs of the users. Economic viability, from the point of view of a developing country, depends on its possessing the requisite knowledge and the capability to participate effectively in a remote sensing enterprise. This often requires long-term intensive education, fostered by, amongst other organisations, the United Nations.

See also *developing countries*.

LEISA (Linear Etalon Imaging Spectrometer Array) U.S. near infrared nadir-viewing imaging spectrometer, carried on *Lewis*. Waveband: 1–2.5 μm. Spectral resolution: 5 nm. Spatial resolution: 300 m. Swath width: 77 km.

LEO See *Low Earth Orbit*.

Lewis U.S. satellite, operated by *NASA*, launched August 1997. Objectives: Proof-of-concept mission for small, low-cost spacecraft. Orbit: Circular *Sun-synchronous LEO* at 523 km altitude; inclination 97.5°. Principal instruments: *HSI*, *LEISA*. The satellite also carries an astronomical instrument.

Note: Control of Lewis was lost shortly after launch.

URL: http://crsphome.ssc.nasa.gov/ssti/welcome.htm

LFC (Large Format Camera) *Photographic system* carried on *Space Shuttle* missions in 1981 and 1984. Focal length: 305 mm. Image format: 230 × 460 mm. Coverage: 170 km × 340 km (from 225 km altitude). Spatial resolution: 10–15 m (from 225 km altitude).

Lidar Acronym for LIght Detection And Ranging (compare *radar*). A remote sensing lidar uses a downward-pointing laser, transmitting very short pulses or a modulated signal in the optical or near infrared part of the electromagnetic spectrum. The backscattered radiation is detected and analysed for time delay, amplitude, polarisation or frequency, depending on the application:

Laser profiler (laser altimeter). The wavelength of the laser radiation is chosen such that atmospheric attenuation is small. The two-way travel time from the instrument to the Earth's surface or cloud-top is used to determine the surface altitude. The horizontal resolution of a laser profiler is determined by the height and by the laser's beamwidth. Since the latter can be as small as a few tenths of a milliradian, resolutions of the order of 100 m are possible from spaceborne laser profilers. The height resolution is determined by the rise-time of the pulse (or equivalently the bandwidth of the modulating signal) and the signal-to-noise ratio of the received signal. Height resolutions of better than 1 m are possible from spaceborne laser profilers (compare *radar altimeter*). The first laser profiler to be operated from a satellite is *Balkan-1*; by 2002 or 2003 both the *Balkan-2* and *GLAS* instruments should be operational.

Backscatter lidar. The operation of a backscatter lidar is similar to that of a laser profiler, but the radiation backscattered from atmospheric constituents such as aerosols and optically thin clouds is used to calculate vertical profiles or integrated values of the backscattering coefficient. The horizontal resolution of a backscatter lidar is similar to that of a laser profiler, but the vertical resolution is poorer (typically 10–200 m). This is because the backscattered signal must be integrated over a range of heights to give a detectable output. The first backscatter lidar to be operated from a satellite is *Alissa*; by 2002 or 2003, both the *Balkan-2* and *GLAS* instruments should be operational.

Differential absorption lidar (DIAL). A differential absorption lidar uses a tuneable laser to measure the spectral variation of the backscattered signal. This adds to the backscatter lidar the ability to measure the profiles of absorbing constituents such as water vapour. Although the differential absorption lidar technique has been verified from airborne platforms, no instruments have yet been placed on satellites.

Wind lidar. A wind lidar measures the Doppler shift of the backscattered signal. This adds to the backscatter lidar the ability to determine the component of the scattering medium's velocity along the line of sight (i.e. the vertical component for a downward-looking lidar). As with the differential absorption lidar, no examples of this type of instrument have yet been placed in space.

Lightning Electrical discharge between charged regions of a thundercloud, or between the cloud and the Earth's surface. Characterisation of the global distribution and variability of lightning can provide improved knowledge of the distribution of *cloud*, and hence input to hydrological models. Lightning can be detected in the optical/near infrared region. The longest-running data set is currently provided by the *OLS* sensor on the *DMSP* satellite programme. A dedicated satellite lightning sensor, *LIS*, is now carried on board *TRMM*.

Lightning Imaging Sensor See *LIS*.

Limb Infrared Monitor of the Stratosphere See *LIMS*.

Limb Radiance and Inversion Radiometer See *LRIR*.

Limb-sounding A technique for atmospheric profiling in which the sensor's line of sight is almost tangential to the Earth's surface (i.e. it passes through the atmosphere's limb). Limb-sounding can be used to measure thermally emitted radiation, or the absorption of radiation originating outside the atmosphere – usually from the Sun or Moon, but also from stars. This geometry gives better vertical resolution (typically 1 or 2 km) than nadir sounding, at the expense of poor horizontal resolution (of the order of 300 km) in the direction of the line of sight.

Limb-sounding of thermally emitted radiation is usually performed at thermal infrared wavelengths (examples: *CLAES, HiRDLS, LIMS, LRIR, MIPAS, SAMS*), but can also be undertaken using short-wavelength passive microwave radiation (examples: *AMAS, MLS, SMR*). At longer wavelengths the larger beamwidth would significantly degrade the vertical resolution of the technique. Emission limb-sounding is used to obtain vertical profiles of atmospheric temperature and concentrations of water vapour, ozone and other molecular species.

Limb-sounding of absorption spectra is performed at thermal infrared wavelengths (e.g. *HALOE, ILAS, Istok-1*) and at ultraviolet, visible and near-infrared wavelengths (e.g. *GOMOS, ILAS, Ozon-M, POAM-2, SAGE-II/III, SAM-II, Sciamachy*). As with emission measurements, absorption measurements (often called occultation measurements) are used to measure temperature, water vapour, ozone and other molecular species in the atmosphere.

Since a satellite in *LEO* makes approximately 14 orbits per day, a solar occultation limb sounder has approximately 28 measurement opportunities (14 sunrises and 14 sunsets) every day. The same is true for lunar occultation. However, stellar occultation observations (e.g. *GOMOS*) can be made anywhere on the Earth's night side.

LIMS (Limb Infrared Monitor of the Stratosphere) U.S. thermal infrared (emission) limb sounder, for vertical profiling of stratospheric temperature, H_2O, NO_2, O_3 and other gases, carried on *Nimbus-7*. <u>Wavebands</u>: 6.25, 6.75, 9.65, 11.35, 15.25, 13.3–17.2 μm. <u>Height resolution</u>: 2 km. <u>Height range</u>: 10–65 km.
LIMS ceased operation in 1979.

Lineaments See *geology*.

Linear enhancement See *contrast enhancement*.

Linear Etalon Imaging Spectrometer Array See *LEISA*.

Linear Imaging Self-Scanner See *LISS*.

Linear polarisation See *polarisation*.

Linear stretch See *contrast enhancement*.

Line detection The process of detecting *pixels* in an image that form part of a linear feature. If the feature is several pixels wide, its two edges can be detected by *edge detection*. If it is only one pixel wide, it can be detected using a set of *convolution operators* designed to match line segments in various orientations. (This is a simple example of *shape detection*.) For example, the following set of kernels will detect, respectively, horizontal, vertical, and the two diagonal orientations:

$$
\begin{array}{ccc}
-1 & -1 & -1 \\
2 & 2 & 2 \\
-1 & -1 & -1
\end{array}
\qquad
\begin{array}{ccc}
-1 & 2 & -1 \\
-1 & 2 & -1 \\
-1 & 2 & -1
\end{array}
$$

$$
\begin{array}{ccc}
-1 & -1 & 2 \\
-1 & 2 & -1 \\
2 & -1 & -1
\end{array}
\qquad
\begin{array}{ccc}
2 & -1 & -1 \\
-1 & 2 & -1 \\
-1 & -1 & 2
\end{array}
$$

In a typical application, the pixel will be judged to form part of a line if the sum of the squares of the outputs from these four operators exceeds some threshold. See also *Hough transform*.

LIS (Lightning Imaging Sensor) U.S. single-waveband near infrared *CCD* (step-stare) imaging radiometer, for observation of lightning. Carried on *TRMM*. Waveband: 777 nm. Spatial resolution: 4 km. Image size: 600 km × 600 km.

URL: http://thunder.msfc.nasa.gov/lis.html

LISS (Linear Imaging Self-Scanning Sensor) Indian optical/near infrared imaging *CCD* (pushbroom) radiometer carried by the *IRS* satellites.

LISS-1 Wavebands: 0.46–0.52, 0.52–0.59, 0.62–0.68, 0.77–0.86 μm. Spatial resolution: 72 m. Swath width: 148 km.

LISS-2 Wavebands: 0.45–0.52, 0.52–0.59, 0.62–0.68, 0.77–0.86 μm. Spatial resolution: 36 m. Swath width: 145 km (as two 74-km swaths with 3 km overlap).

LISS-3 Wavebands: 0.52–0.59, 0.62–0.68, 0.77–0.86, 1.55–1.75 μm. Spatial resolution: 24 m (71 m for 1.55–1.75 μm band). Swath width: 142 km.

LISS-4 (proposed). Wavebands: optical, near infrared. Spatial resolution: approximately 10 m. Swath width: 40 km.

Local area coverage See *AVHRR*.

Local crossing time The time at which a satellite crosses a particular latitude, especially the equator, in a specified sense (either northbound or southbound). For a satellite in a *Sun-synchronous orbit* this time depends only on the latitude and the sense of the crossing (i.e. it is independent of the longitude and the date), if due allowance is made for any differences between the *mean local solar time* and the zone time.

Logarithmic contrast enhancement See *contrast enhancement*.

Long-wave radiation In climatology, long-wave radiation is thermal infrared radiation having a maximum *spectral radiance* at about 10 µm.

Look direction The direction of view of a sensor (normally of an *active* system) when gathering information from a particular area on the Earth's surface or in its atmosphere.

Looks Multi-look processing is a technique for improving the *radiometric resolution* of a *synthetic aperture radar* image at the expense of the *spatial resolution*. The processing is performed at the stage of focussing the image, i.e. of combining the detected amplitudes and phases to form the image. The available bandwidth in range or azimuth is divided into a number N (termed the number of looks) of non-overlapping sub-bands, each of which is processed separately. The resulting images are combined incoherently to produce an N-look image.
See also *speckle*.

Low Earth Orbit (LEO) A satellite orbit with a *semimajor axis* significantly less than that of a *geostationary orbit*. In practice, an orbit with a minimum height of between approximately 600 and 2000 km above the Earth's surface.

Low-pass filter See *smoothing*.

LOWTRAN (LOW resolution TRANsmittance) A computer model of the *radiative transfer equation* in the Earth's atmosphere, used for *atmospheric correction*. The earliest LOWTRAN model covered the wavelength range from 0.25 µm to 28.5 µm in steps of $5\,\text{cm}^{-1}$. More recent versions include improved models of *aerosols*, *water vapour*, *cloud* and *rain*, and also cover the microwave region of the electromagnetic spectrum. The MODTRAN model also has an increased spectral resolution of $2\,\text{cm}^{-1}$.

LRIR (Limb Radiance and Inversion Radiometer) U.S. thermal infrared (emission) limb sounder, for vertical profiling of atmospheric ozone, carried on *Nimbus-6*. Waveband: 4 channels between 9.6 and 37 µm.

Lumen The unit of luminous flux. See *photometric quantities*.

Lux The unit of illuminance, equal to one *lumen* per square metre. See *photometric quantities*.

Magnetic permeability See *permeability*.

Mahalanobis distance See *supervised classification*.

Main lobe See *power pattern*.

Major axis See *orbit*.

Map projection The geometric transformation between the geographical coordinates (latitude ϕ and longitude λ) of a point on the Earth's surface, and the Cartesian coordinates (x, y) on a map or in an image. Since the Earth's surface is curved in two dimensions, it cannot be projected onto a plane without introducing some distortions (a *scale* that varies with position and orientation), and in consequence a large number of projections have been developed to minimise the effects of these distortions in various ways.

Some of the commonest simple projections are defined below. The formulae assume that ϕ and λ are measured in radians, that $\phi = -\pi/2$ at the South Pole, 0 at the equator and $+\pi/2$ at the North Pole, that λ increases from west to east, that the Earth is a sphere of radius R, and that the principal scale of the map is s (<1). For the Transverse Mercator projection, λ_0 is the longitude of the central meridian. The projection is useful only for longitudes within a few degrees of this value. For the Polar Stereographic projection, $\chi = (\pi/2 - \phi)$ for the north polar aspect and $(\phi + \pi/2)$ for the south polar aspect. Note that offsets are often added to the resultant values of x and y to bring them into a convenient range, and that a constant scale factor is often applied to the whole map so that the spatially averaged scale is correct.

Lambert equal-area

$$x = Rs\lambda,$$

$$y = Rs\sin\phi.$$

Sinusoidal

$$x = Rs\lambda\cos\phi,$$

$$y = Rs\phi.$$

Mercator

$$x = Rs\lambda,$$

$$y = Rs\ln\tan\left(\frac{\pi}{4} + \frac{\phi}{2}\right).$$

Transverse Mercator

$$x = Rs\ln\tan\left(\frac{\pi}{4} + \frac{1}{2}\arcsin(\sin(\lambda - \lambda_0)\cos\phi)\right),$$

$$y = Rs\arctan\left(\frac{\tan\phi}{\cos(\lambda - \lambda_0)}\right).$$

Polar Stereographic

$$x = 2Rs\tan\frac{\chi}{2}\cos\lambda,$$

$$y = 2Rs\tan\frac{\chi}{2}\sin\lambda.$$

For large-scale maps (1/25 000 or larger) the effects of the Earth's curvature are insignificant, except in so far as they determine the length of a degree of latitude, and the choice of map projection is fairly arbitrary. For very small-scale maps (1/5 000 000 or smaller) the assumption of a spherical Earth is sufficiently accurate, and the above formulae may be used without modification. However, for maps of intermediate scales the Earth's asphericity introduces significant errors unless allowance is made for it. This is normally done by assuming that the Earth is an *ellipsoid* of revolution, rather than a sphere, characterised by its equatorial axis a and its polar axis b. This considerably increases the complexity of the projection formulae.

Marine Observation Satellite See *MOS-1*.

Maximum likelihood classification See *supervised classification*.

Maxwell's equations The fundamental equations governing the relationship between electric and magnetic fields, and hence the propagation of *electromagnetic radiation*. The equations can be written as

$$\nabla \cdot \mathbf{D} = \rho,$$

$$\nabla \cdot \mathbf{B} = 0,$$

$$\nabla \times \mathbf{E} = -\dot{\mathbf{B}},$$

$$\nabla \times \mathbf{H} = \dot{\mathbf{D}} + \mathbf{j},$$

where $\mathbf{D}$ is the displacement current, $\mathbf{B}$ is the magnetic flux density, $\mathbf{E}$ is the electric field strength, $\mathbf{H}$ is the magnetic field strength, ρ is the volume density of charge and $\mathbf{j}$ is the current density.

MC (Metric Camera) *Photographic system* carried on *Spacelab-1*. Focal length: 305 mm. Image format: 230 × 230 mm. Coverage: 190 km × 190 km. Spatial resolution: 12 m.

Mean local solar time The mean local solar time at a place whose longitude is L degrees east of the Greenwich meridian is

$$T + (L/15) \text{ hours,}$$

where T is the Greenwich Mean Time (GMT). Note that the mean local solar time is likely to be different from the 'official' local time (zone time), since the zone time usually differs from GMT by an integral number of hours. For a *Sun-synchronous orbit* with *inclination i*, the mean local solar time (in hours) at latitude B is given by the following expressions:

	northbound	southbound
N hemisphere	$T_0 - A$	$T_0 + A - 12$
S hemisphere	$T_0 + A$	$T_0 + 12 - A$

where T_0 is the mean local solar time, in hours, for a northbound equator crossing, and

$$A = \frac{12}{\pi} \arccos \sqrt{\sec^2 B - \frac{\tan^2 B}{\sin^2 i}}.$$

The inverse cosine is evaluated in radians.
See also *solar illumination direction*.

Measurement of Pollution in the Troposphere See *MOPITT*.

Median filter See *smoothing*.

Medium Resolution Imaging Spectrometer See *MERIS*.

Medium Resolution Infrared Radiometer See *MRIR*.

Mercator projection See *map projection*.

MERIS (Medium Resolution Imaging Spectrometer) European optical/near infrared *CCD* (pushbroom) imaging spectrometer, to be included on the *Envisat* satellite. Wavebands: 15 bands between 0.40 and 1.05 μm. The band centres are programmable, as are the bandwidths which can be set between 2.5 and 25 nm. Spatial resolution: 250 m or 1 km. Swath width: 1500 km.

1 km resolution imagery can be obtained continuously, 250 m resolution imagery for approximately 10 minutes per orbit.

URL: http://envisat.estec.esa.nl/instruments/meris/index.html

Mesosphere Region of the Earth's *atmosphere*, lying between heights of approximately 50 and 85 km.

MESSR (Multispectral Electronic Self-Scanning Radiometer) Japanese optical/near infrared mechanically scanned imaging radiometer, carried on *MOS-1*

satellites. <u>Wavebands</u>: 0.51–0.59, 0.61–0.69, 0.73–0.80, 0.80–1.10 μm. <u>Spatial resolution</u>: 50 m. <u>Swath width</u>: 100 km × 2 (two instruments carried).

URL: http://hdsn.eoc.nasda.go.jp/guide/guide/satellite/sendata/messr_e.html

Meteor-1 Soviet satellite series. First satellite launched 1969; last in series 1977. <u>Objectives</u>: Meteorological monitoring. <u>Orbit</u>: Circular *LEO* at 650 km altitude (to 1971), 900 km from 1971. Period 98 minutes (to 1971), 103 minutes (from 1971); inclination 81°. <u>Principal instruments</u>: *AC, IR, TV*.

The satellites were launched so as to keep two or three operational at any one time.

Meteor-2 Soviet/Russian satellite series. First satellite launched 1975; programme continuing. <u>Objectives</u>: Meteorological monitoring. <u>Orbit</u>: Circular *LEO* at 800 to 1000 km altitude. Period 101 to 104 minutes; inclination 81° to 83°. <u>Principal instruments</u>: *IR, SM, TV*. The satellites also carry solar–terrestrial physics packages.

Meteor-3 Soviet/Russian satellite series. First satellite launched 1985; programme continuing. <u>Objectives</u>: Meteorological and environmental monitoring. <u>Orbit</u>: Circular *LEO* at 1230 km altitude. Period 109 minutes; inclination 82.5°. <u>Principal instruments</u>: *IR* (up to and including Meteor-3-6, launched 1991), *Klimat* (Meteor-3-3, launched 1988, onward), *MR-2000M, MR-900B, Scarab* (Meteor-3-7, launched 1994), *SM, TOMS* (Meteor-3-6). Meteor-3-7 also carries the *PRARE* ranging system.

Meteor-3M Russian satellite series, planned to begin with the launch of Meteor-3M-1 in 1999. <u>Objectives</u>: to combine the meteorological programme of the earlier Meteor series with the surface-observation programme of the *Resurs* programme. <u>Orbit</u>: Circular *Sun-synchronous LEO* at 1020 km altitude. Period 105 minutes; inclination 99.5°; equator-crossing time 10:30 (ascending). <u>Principal instruments</u>: *AMAS, BUFS-4, MIVZA, MTZA, MVZA, MZOAS, SAGE III, SFOR-1*.

The inclusion of the *SAGE III* instrument on Meteor-3M-1 provides the function originally planned for the **EOS-Aero** satellite as part of the *EOS* programme.

Meteorology applications Observations from satellite sensors are of great value in both operational meteorology and research investigations; information can be obtained from remote areas of the Earth where there are few *in situ* observations. For routine weather analysis, satellite imagery can show the locations of weather systems through their associated cloud, as well as isolated fronts and troughs. Imagery is also very important for identifying mesoscale (less than 1000 km horizontal diameter) weather systems that may have a major effect on surface conditions, but which may be too small to be detected by the surface observing network. Imagery is also very valuable in predicting conditions, especially for the period up to 12 hours ahead (nowcasting). Here sequences of images may be displayed as a 'movie-loop' to show the speed and direction of movement of cloud elements and weather systems as a whole. Some objective data may be determined from a series of satellite images, such as cloud drift winds, provided that suitable tracers can be found. However, for numerical weather prediction, the most important forms of satellite data are the atmospheric temperature soundings obtained from infrared and microwave radiometers. See *atmospheric sounding, climate, clouds, precipitation, temperature sounding, water vapour, wind speed and velocity.*

Meteor-Priroda Soviet/Russian satellite series. First satellite launched 1985; programme continuing. Objectives: Meteorological and environmental monitoring. Orbit: Circular *Sun-synchronous LEO* at 620 km altitude. Period 92.7 minutes; inclination 97.9°. *Exactly repeating orbit* (237 orbits in 16 days). Principal instruments: *Fragment-2* (Meteor-Priroda-5, launched 1980, only), *IR* (up to Meteor-Priroda-2, launched 1976), *MSU-E, MSU-M, MSU-S, MSU-SK* (Meteor-Priroda-5 only), *SHF, SI-GDR* (Meteor-Priroda-2 to 4, launched 1979), *TV* (Meteor-Priroda-1 only).

Most of the series also carried solar–terrestrial physics packages. The series was a prototype for the *Resurs-O1* series.

Meteosat European satellite series, operated by *ESA* (to 1988) and by *EUMET-SAT* after 1988. The first satellite in the series was launched in November 1977. Objectives: Operational meteorological observations, collection and relay of weather data. Orbit: *Geostationary* (longitude 0°). Principal instruments: *VISSR*. Meteosat satellites also carry a data collection package (similar in function to *Argos*).

The Meteosat programme is now known as MOP (Meteosat Operational Programme). From about the year 2000, the programme will be upgraded to the *MSG* (Meteosat second generation) series.

URL: http://www.esoc.esa.de/external/mso/meteosat.html

Methane A significant radiatively active component of the Earth's atmosphere. Chemical formula CH_4. The principal atmospheric absorption lines of methane are at 1.66, 2.1, 2.3 and 2.4 μm, with a broad line at about 3.3 μm.

Metop (Meteorology Operational Programme) European satellite programme operated by EUMETSAT as part of the *POEM* mission. Metop-1 is scheduled for launch in 2000 with a nominal lifetime of 4 years. Objectives: Operational meteorological observations, providing continuity with *NOAA* satellite missions. Orbit: Circular *Sun-synchronous LEO* at 850 km altitude. Period 102 minutes; inclination 99°; equator crossing time 09:00. Principal instruments: *AMSU-A/ MHS, ASCAT, AVHRR/3, HIRS 3, IASI, OMI, Scarab*. The satellite will also carry the *Argos* data collection package, as well as solar–terrestrial physics and search and rescue packages.

URL: http://www.esrin.esa.it/esa/progs/METOP.html

Metric Camera See *MC*.

MHS See *AMSU/MHS*.

Michelson Interferometric Passive Atmospheric Sounder See *MIPAS*.

MicroMAPS (Micro-Measurement of Air Pollution from Satellites) Canadian gas-cell spectrometer for monitoring atmospheric CO and N_2O, planned for inclusion on *Clark* satellite. Waveband: 4.67 µm.

Microwave Subdivision of the *electromagnetic spectrum*, corresponding to frequencies between 0.3 and 300 GHz (wavelengths between 1 mm and 1 m). Both *active* and *passive* remote sensing is performed at microwave frequencies: see *passive microwave radiometry, radar*.

Microwave Humidity Sounder See *AMSU/MHS*.

Microwave Limb Sounder See *MLS*.

Microwave radiometer See *MWR*.

Microwave Scanning Radiometer See *MSR*.

Microwave sounding instruments Microwave sounding instruments are used to determine the profiles of atmospheric temperature and humidity for operational and research meteorological applications. They have the advantage over infrared sounders of being able to provide profiles through the whole depth of the atmosphere in cloudy conditions. The Microwave Sounding Unit (*MSU*) was an early operational instrument with a relatively coarse footprint of close to 150 km at nadir. It suffered from only having five channels with broad weighting functions which resulted in little vertical structure in the temperature profiles. The Advanced Microwave Sounder Unit (*AMSU*) has 20 channels and can be used for temperature and humidity sounding. The first 15 channels (AMSU-A) will be used for temperature sounding and have a horizontal resolution of 40 km at nadir. Channels 3 to 14 of AMSU-A are located in the 50–60 GHz oxygen resonance band. Channels 16 to 20 of AMSU (AMSU-B) will be used for humidity sounding and to detect precipitation under clouds and will consist of one channel at 89 GHz and four on the wings of the water vapour resonance line at 183.3 GHz. On later EOS satellites AMSU-B will be known as the Microwave Humidity Sounder (MHS). AMSU-B/MHS will produce soundings with a horizontal resolution of 15 km.

Microwave Sounding Unit See *MSU*.

Middle infrared See *infrared*.

Midori See *ADEOS*.

Mie scattering *Scattering* of radiation by particles larger than about $\lambda/100$, where λ is the wavelength of the radiation. For spherical particles, the scattering and absorption cross-sections σ_S and σ_A, defined as the power scattered and absorbed divided by the incident flux, can be expressed as a power series in the dimensionless variable $x = 2\pi a/\lambda$, where a is the particle radius:

$$\sigma_S = \pi a^2 \sum_{m=1}^{\infty} \alpha_m x^m,$$

$$\sigma_A = \pi a^2 \sum_{m=1}^{\infty} \beta_m x^m.$$

If the particle has *refractive index* n, the first few coefficients are given by

$$\alpha_1 = \alpha_2 = \alpha_3 = 0$$

$$\alpha_4 = \frac{8}{3} \left| \frac{n^2 - 1}{n^2 + 2} \right|^2,$$

$$\alpha_5 = 0,$$

$$\alpha_6 = \frac{16}{45} \left| \frac{(n^2 - 1)^2 (n^2 - 2)}{(n^2 + 2)^3} \right|,$$

$$\alpha_7 = \frac{32}{27} \left| \frac{n^2 - 1}{n^2 + 2} \right|^3,$$

$$\beta_1 = -4 \,\mathrm{Im} \left(\frac{n^2 - 1}{n^2 + 2} \right),$$

$$\beta_2 = 0,$$

$$\beta_3 = -\frac{12}{5} \,\mathrm{Im} \left(\frac{(n^2 - 1)(n^2 - 2)}{(n^2 + 2)^2} \right) + \frac{3}{2\pi} \,\mathrm{Re}(n^2 - 1),$$

$$\beta_4 = 0,$$

$$\beta_5 = \frac{8}{3} \,\mathrm{Re} \left(\frac{n^2 - 1}{n^2 + 2} \right)^2 - \frac{2}{3} \,\mathrm{Im} \left(\frac{n^2 - 1}{2n^2 + 3} \right).$$

Rayleigh scattering describes the situation in which all terms except the first non-zero terms can be ignored.

MIMR (Multifrequency Imaging Microwave Radiometer) European mechanically scanned passive microwave radiometer, planned for inclusion on *EOS PM*. Frequencies: 6.8, 10.65, 18.7, 23.8, 36.5, 89 GHz. Polarisation: H and V. Spatial resolution: 88 km at 6.8 GHz to 7 km at 89 GHz (from Metop). Swath width: 1600 km (from Metop). Sensitivity: 0.2 K at 6.8 GHz to 0.7 K at 89 GHz. Absolute accuracy: 1 K at 6.8 GHz to 1.5 K at 89 GHz.

Minor axis See *orbit*.

MIPAS (Michelson Interferometer for Passive Atmospheric Sounding) European thermal infrared (emission) limb sounder, for measuring a wide range of atmospheric trace gases, planned for inclusion on *Envisat*. Waveband: 4.15 to 14.6 μm. Spectral resolution: selectable between 1 and 10 GHz (Michelson interferometer). Spatial resolution: 300 km (along track), 30 km (across track), 3 km (vertical). Height range: 5–100 km.

URL: http://envisat.estec.esa.nl/instruments/mipas/index.html

Mir-1 Russian space station, launched February 1986. Remote sensing modules have been added to the station at various times. Objectives: Earth observation, amongst others. Orbit: Nominally circular, at 350 to 410 km altitude. Period 92 minutes; inclination 51.6°. Principal instruments on core module: *MKS-M-AS*, *MKS-M2-BS*. Principal instruments on Kristall module: *KFA-1000, Priroda-5*. Principal instruments on Kvant-2 module: *ITS-7D, MKF-6MA, MKS-M2-AS, MKS-M2-BS, KAP-350, KL-103W, Phase*. Principal instruments on Priroda module: *Alissa, Greben, IKAR-Delta, IKAR-N, IKAR-P, Istok-1, MOS-A, MOS-B, MSU-SK, MSU-E, Ozon-M, R400, Travers*. Principal instruments on Spektr module: *Balkan-1, Infrared radiometers, KR-05*.

The Spektr module was added in May 1995, the Priroda module in April 1996.

URL: http://www.hq.nasa.gov/osf/mir/
http://www.ire.rssi.ru/priroda/priroda.htm

MISR (Multi-angle imaging spectroradiometer) U.S. optical/near infrared *CCD* imaging radiometer, planned for inclusion on the *EOS-AM* satellites. Wavelengths: 443, 555, 670, 865 nm. Spatial resolution: 275, 550 or 1100 m (programmable). Swath width: 360 km. Each spectral band is observed in nine different directions: fore, aft and nadir, with views to each side and towards the sub-satellite track.

The MISR will provide measurements of planetary albedo, aerosols, atmospheric scattering, and vegetation parameters. It will give complete global coverage in 9 days (2 days for polar regions).

URL: http://www-misr.jpl.nasa.gov/

Mission to Planet Earth (MTPE) A long-term U.S. programme of spaceborne and other observations of the Earth and its interactions, particularly hydrology, biogeochemistry, and atmospheric, ecological and geophysical processes. The main element of the MTPE is the *EOS* programme, but other U.S. and non-U.S. satellite missions and programmes are increasingly being integrated within the MTPE framework.

URL: http://www.hq.nasa.gov/office/mtpe/

MIVZA Russian conically scanning passive microwave radiometer, planned for inclusion on *Meteor-3M* satellites. Frequencies: 20, 35, 94 GHz. Bandwidth:

2.4 GHz (all channels). Polarisation: H and V (94 GHz H only). Spatial resolution: 100, 60, 50 km at nadir. Incidence angle: 42°. Swath width: 1350 km. MIVZA observations will be used for estimating total atmospheric water vapour and rain rate, land ice and sea ice monitoring, and wind speed over oceans.

MK4 *Photographic system* carried on *Resurs-F*. Focal length: 300 mm. Format: 180 × 180 mm. Wavebands: 0.44–0.68, 0.46–0.51, 0.52–0.57, 0.64–0.69, 0.61–0.75, 0.81–0.86 μm. Coverage: 144 × 144 km (from 240 km altitude). Spatial resolution: 10–14 m (from 240 km altitude) depending on film type.

MK4 has four lenses and can acquire images in four of the following six spectral bands, defined by filters: 0.44–0.68, 0.46–0.51, 0.61–0.75, 0.64–0.69, 0.52–0.57, 0.81–0.86 μm. The MK-4M system has just the last four of these bands.

URL: http://www.eurimage.it/Products/MK_4.html

MKF-6 *Photographic system* carried on *Salyut* space station missions 6 and 7. Focal length: 125 mm. Wavebands: 0.46–0.50, 0.52–0.56, 0.58–0.62, 0.64–0.68, 0.70–0.74, 0.79–0.89 μm. Coverage: 180 km from 250 km altitude.

MKF-6MA *Photographic system* carried on *Mir-1* (Kvant-2 module). Focal length: 135 mm. Waveband: 0.48–0.84 μm (filter). Format: 81 × 56 mm. Coverage: 145 × 210 km (from 350 km). Spatial resolution: 30 m (from 350 km).

MKS-M-AS (Multichannel Atmospheric Spectrometer) German nadir-viewing near infrared spectrometer, carried on *Mir-1* core and (as MKS-M2-AS) Kvant-2 modules. The latter module is steerable. Waveband: 757–770 nm. Spectral resolution: 1.5 nm. Spatial resolution: 6.3 × 0.5 km.

The instrument provides atmospheric temperature profiles for atmospheric temperature correction.

MKS-M2-BS (Multichannel Biospectrometer) German optical/near infrared non-imaging spectrometer, carried on core and Kvant-2 modules of *Mir-1* space station. Wavebands: 415–830 nm (415–1030 nm on Kvant-2), spectral resolution 10 nm. Spatial resolution: 2.4 km.

MLS (Microwave Limb Sounder) U.S. passive microwave (emission) limb sounder, for vertical profiling of atmospheric pressure (from O_2 concentration), temperature, H_2O, O_3, ClO and other trace gases, carried on *UARS* and planned for inclusion on *EOS-Chem*. Wavebands: UARS: 63 GHz, 183 GHz (2 channels), 205 GHz (3 channels); EOS: 215, 310, 640, 2500 GHz. Observation range: 0–120 km. Spatial resolution: UARS: 400 km (along-track), 4 km (across-track), 4 km (vertical); EOS: 300 km (along-track), 3 km (across-track), 1.2 km (vertical).

Moderate Resolution Imaging Spectrometer See *MODIS(-N)*.

MODIS(-N) (Moderate Resolution Imaging Spectrometer [-Nadir]) U.S. optical/ near infrared/thermal infrared mechanically scanned *CCD* imaging spectrometer, planned for inclusion on *EOS-AM* and *EOS-PM* satellites. Wavebands: 36 bands between 0.4 and 15.0 μm, defined by interference filters. The visible-wavelength bands include narrow (10 nm) bandwidths suitable for ocean colour observations. Spatial resolution: 250, 500 and 1000 m. Swath width: 2300 km.

URL: http://ltpwww.gsfc.nasa.gov/MODIS/MODIS.html

MODTRAN See *LOWTRAN*.

Modulation transfer function The modulation transfer function (MTF) provides a more detailed representation of the spatial response of an imaging system than is given by a single parameter such as the *spatial resolution*. The concept is based on the principle of decomposing spatial variations in the brightness of the target area and the detected image into spatial frequencies. The MTF is then defined, as a function of spatial frequency, as the ratio of the amplitude of a particular spatial frequency in the image to the amplitude of the corresponding spatial frequency in the target.

The usual single-parameter measures of spatial resolution can be retrieved from the MTF. One advantage of the MTF is that it is often possible to define it separately for each component of an imaging system, including the atmospheric transmission, such that the overall MTF of the system is the product of the MTFs of each component.

MOMO See *MOS-1*.

Monostatic See *radar equation*.

MOP See *Meteosat*.

MOPITT (Measurement of pollution in the troposphere) Canadian scanning infrared correlation spectrometer, planned for inclusion on *EOS-AM-1* satellite. Wavebands: 2.3, 2.4 and 2.7 μm (gas cell spectrometer). Spatial resolution: 22 km. Swath width: 640 km.

MOPITT will be used to measure carbon dioxide profiles (vertical resolution 4 km) and total methane content of the atmosphere.

URL: http://eos.acd.ucar.edu/mopitt/home.html

MOS (Multispectral Optoelectronic Scanner [Modular Optoelectronic Scanning Spectrometer]) German optical/near infrared mechanically scanned imaging spectrometer, carried on *Mir-1* space station (MOS-A and -B only) and *IRS-P3*.
MOS-A Wavebands: 756.7, 760.6, 763.5, 766.4 nm (bandwidth 1.4 nm). Spatial resolution: 2.8 km (Mir), 5.8 km (IRS-P3). Swath width: 80 km (Mir), 200 km (IRS-P2).

MOS-B Wavebands: 408, 443, 485, 520, 570, 615, 650, 685, 750, 815, 870, 945, 1010 nm (bandwidth 10 nm). Spatial resolution: 0.7 km (Mir), 1.5 km (IRS-P2). Swath width: 80 km (Mir), 200 km (IRS-P3).

MOS-C Wavebands: 2 bands between 1.6 and 2.3 μm. Spatial resolution: 1.5 km. Swath width: 200 km.

The MOS instrument uses a diffraction-grating spectrometer. Its principal application is to measurements of ocean colour, aerosol and cloud characteristics.

MOS-1 (Marine Observation Satellite) Japanese satellite series, operated by *NASDA*. MOS-1A was launched in February 1987, MOS-1B in February 1990. The programme was terminated in April 1996. Objectives: Ocean colour, atmospheric water vapour and land surface observations. (Similar function to *Landsat*.) Orbit: Circular *Sun-synchronous LEO* at 908 km altitude. Period 103.3 minutes; inclination 99.1°; equator crossing time 10:15 to 10:30. *Exactly-repeating orbit* (237 orbits in 17 days). Principal instruments: *MESSR, MSR, VTIR*. The satellites also carry data collection packages.

MOS-1 satellites are also known as MOMO.

URL: http://hdsn.eoc.nasda.go.jp/guide/guide/satellite/satdata/mos_e.html

Mosaic An assemblage of contiguous or overlapping images to provide coverage of a region larger than can be encompassed by any one image. Effective mosaicking requires very accurate *radiometric correction* and *geometric correction* of the constituent images.

MR-2000 Soviet/Russian *vidicon* visible-wavelength imaging radiometer, carried on *METEOR-2* satellites. Waveband: 0.5–0.8 μm. Spatial resolution: 1 km at nadir. Swath width: 2600 km.

MR-2000M Soviet/Russian *vidicon* visible-wavelength imaging radiometer, carried on *METEOR-3* satellites. Waveband: 0.5–0.7 μm. Spatial resolution: 700 m at nadir. Swath width: 3100 km.

On-board data storage, for downlinking to receiving station.

MR-900B Soviet/Russian *vidicon* visible-wavelength imaging radiometer, carried on *METEOR-3* satellites. Waveband: 0.5–0.7 μm. Spatial resolution: 1 km at nadir. Swath width: 2600 km.

MRIR (Medium Resolution Infrared Radiometer) U.S. ultraviolet/optical/infrared nadir-viewing spectrometer, for radiation budget and surface albedo measurements, carried on *Nimbus-2* and *-3* satellites. Wavebands: 5 channels between 0.2 and 30 μm. Spatial resolution: 55 km.

MSG (Meteosat second generation) European satellite series, operated by *EUMETSAT*, planned to begin in 2000. Objectives: Operational meteorological observations, collection and relay of weather data. MSG is an enhanced version of the *Meteosat* programme. Orbit: *Geostationary* (longitude 0°). Principal instruments: *SEVIRI*. MSG satellites also carry a data collection package (similar in function to *Argos*) and a search and rescue system.

URL: http://www.esoc.esa.de/external/mso/meteosat.html

MSR

1. See *VISSR*.
2. (Microwave Scanning Radiometer) Japanese scanning passive microwave radiometer, carried on *MOS-1* satellites. Frequencies: 23.8 GHz (bandwidth 400 MHz), 31.4 GHz (bandwidth 500 MHz). Spatial resolution: 32 km, 23 km, respectively. Swath width: 320 km. Absolute accuracy: 1.5 K.

MSS (Multispectral Scanner) U.S. optical/near infrared/thermal infrared mechanically scanned imaging radiometer carried on *Landsat-1* to *-5* satellites.

Wavebands:

	Band number	
μm	Landsat-1 to -3	Landsat-4, -5
0.50–0.60	4	1
0.60–0.70	5	2
0.70–0.80	6	3
0.80–1.10	7	4
10.4–12.6	8 (Landsat-3 only; March to July 1978)	

Spatial resolution: 79 m. Swath width: 185 km.

The numbering of the bands on the Landsat-1 to -3 instruments was adopted to avoid confusion with the bands of the *RBV* instruments also carried by those satellites.

URL: http://edcwww.cr.usgs.gov/glis/hyper/guide/landsat

MSU (Microwave Sounding Unit) U.S. scanning passive microwave radiometer for atmospheric temperature profiling, carried on *NOAA-6* onwards, as part of the *TOVS* package, and *TIROS-N*. Wavebands: 50.3, 53.7, 55.0, 58.0 GHz. Spatial resolution: 100 km at nadir. Swath width: 2400 km. Height range: 0–20 km.

MSU-E Soviet/Russian optical/near infrared imaging radiometer on *Almaz-1B*, *Meteor-Priroda*, *Mir-1* and *Resurs-O* satellite series. Wavebands: 0.50–0.60, 0.60–0.70, 0.80–0.90 μm. Spatial resolution: 45 m from 650 km altitude. Swath width: 45 km (80 km for two instruments), steerable to ±300 km from the sub-satellite track.

Resurs-O2 satellites carry the MSU-E1 instrument with a higher spatial resolution of 25 m.

MSU-M Soviet/Russian optical/near infrared mechanically scanned imaging radiometer, carried on *Meteor-Priroda* and *Okean-O1* satellites. Wavebands: 0.5–0.6, 0.6–0.7, 0.7–0.8, 0.8–1.0 µm. Spatial resolution: 1.0 × 1.7 km. Swath width: 2000 km.

MSU-S Soviet/Russian optical/near infrared mechanically scanned imaging radiometer, carried on *Meteor-Priroda* and *Okean-O1* satellites. Wavebands: 0.58–0.70, 0.70–1.0 µm. Spatial resolution: 140 m at nadir (Meteor), 250 m at nadir (Okean). Swath width: 1400 km (Meteor), 1100 km (Okean).

MSU-SK Soviet/Russian optical/near infrared/thermal infrared mechanically scanned imaging radiometer, carried on *Almaz-1B*, *Meteor-Priroda-5*, *Mir-1* and *Resurs-O* satellites, and planned for *Okean-O* satellite. Wavebands: 0.50–0.60, 0.60–0.70, 0.70–0.80, 0.80–1.10, 10.4–12.6 µm. Spatial resolution: 80 m (300 m for thermal infrared band) from Almaz and Mir orbits; twice as large from the higher orbits. Swath width: 300 km from Almaz and Mir orbits, 600 km from the higher orbits.

MSU-V Soviet/Russian optical/near infrared/thermal infrared *CCD* (pushbroom) imaging radiometer, planned for inclusion on *Okean-O* satellite series. Wavebands: 1: 0.45–0.52 µm; 2: 0.52–0.62 µm; 3: 0.62–0.74 µm; 4: 0.76–0.90 µm; 5: 0.90–1.10 µm; 6: 1.55–1.75 µm; 7: 2.10–2.35 µm; 8: 10.3–12.6 µm. Spatial resolution: 50 m (bands 1–5), 100 m (band 6), 275 m (bands 7–8). Swath width: 180 km.

MTF See *modulation transfer function*.

MTPE See *Mission to Planet Earth*.

MTZA Russian conically scanned passive microwave radiometer for atmospheric temperature sounding, planned for inclusion on *Meteor-3M* satellites.

Frequency (GHz)	Bandwidth (MHz)	Spatial resolution (km)
18.7	1000	90
36.5	1000	50
52.2	400	35
55.4	400	35
56.968 ± 0.100*	50	35
56.968 ± 0.050*	20	35
56.968 ± 0.025*	10	35
56.968 ± 0.010*	5	35
56.968 ± 0.005*	3	35
90.5	1000	20

*Tuneable bands

Incidence angle: 42°. Swath width: 1500 km. Temperature sensitivity: 0.05–3.0 K.

Multi-angle Imaging Spectroradiometer See *MISR*.

Multifrequency Imaging Microwave Radiometer See *MIMR*.

Multi-look image See *looks*.

Multispectral Electronic Self-Scanning Radiometer See *MESSR*.

Multispectral imager A type of *electro-optical sensor*, operating in the optical or infrared part of the electromagnetic spectrum and providing radiance data for several (possibly contiguous) spectral bands for each *pixel* of the image. If the number of spectral bands is very small (one or two), the instrument can instead be termed an *imaging radiometer*; if it is large, with contiguous bands, the instrument is generally referred to as a spectrometer, or an imaging spectrometer if it also provides two-dimensional coverage. Various methods are employed to obtain the two-dimensional coverage needed for an imaging system (see *pushbroom scanner, scanning, spin-scan imager, step-stare imager, vidicon, whiskbroom scanner*).

Data from multispectral imagers still provide the bulk of data used in remote sensing applications, partly as a result of their inherent interpretability by comparatively inexpert users. Instruments can conveniently be classified according to their spatial resolution and the part of the electromagnetic spectrum in which they operate. Multispectral imagers offering the highest spatial resolutions tend to have correspondingly limited spatial coverage, so there continues to be a need for lower-resolution sensors offering regional or near-global coverage.

High spatial resolution (≤ 50 m) sensors, with swath widths typically between 30 and 200 km, mostly optical band, predominantly used for land mapping and cartography, agriculture, geological applications, monitoring inland water bodies (e.g. for flood monitoring), coastal zone and cloud observations, and for determining *aerosol* distributions over oceans. Examples: *ASTER, AVNIR, ETM, HRCC, HRG, HRV, HRVIR, LISS, MSU-E, OPS, TM*. Very-high spatial resolution sensors, such as *Carterra-1, EarthWatch Imager, OEA, Orbview 3* and *QuickBird* with resolution finer than 5 m, should be operational by 1999 or 2000. *EarlyBird*, which should have provided 3-m panchromatic resolution, failed shortly after launch.

Intermediate spatial resolution (50–500 m) sensors generally fall into two categories. The first is similar to the high-resolution sensors, but with swath widths typically between 100 and 300 km. Examples: *Fragment-2, IR-MSS, LISS, MESSR, MSS, MSU-SK, MSU-V*. The second contains instruments with narrower bandwidths (generally less than 20 nm), optimised for *ocean colour, aerosol* or *vegetation* observations. These sensors have larger swath widths (typically 400 to 2000 km). Examples: *MERIS, MISR, MODIS-N*.

Low-resolution (≥ 500 m) sensors show greater diversity of application. The following main classes of instrument can be distinguished:

1. Meteorological spin-scan imagers carried by *geostationary* satellites, giving useful coverage of approximately one quarter of the Earth's disc, with a scan time of typically 30 minutes. These instruments normally provide both (broad-band) visible-wavelength and infrared images, with applications to cloud, surface temperature and water vapour monitoring. Recent developments include the ability to measure ocean colour. Examples: *GOES Imager, MVISR, Scanning Radiometer, SEVIRI, VAS, VHRR, VHRSR, VISSR.*

2. Instruments designed primarily to make measurements related to the global radiation budget, especially albedo. These have very wide (in some cases limb-to-limb) swaths, and usually only a few very broad, but accurately calibrated, spectral bands. Examples: *CERES, ERB, MRIR, Scarab.*

3. General-purpose imagers, with bands covering the visible, near infrared and thermal infrared parts of the spectrum, and swath widths ranging typically from 500 to 3000 km. These instruments provide coarse-resolution, wide-swath imaging together with surface and cloud-top temperature data, and find a very wide range of applications over land, oceans and the cryosphere. Over land, applications include measurement of albedo, surface temperature, vegetation distribution and characteristics (including agricultural and forestry applications), monitoring flooding, forest fires, volcanoes and drought. Over oceans, applications include delineation and monitoring of large-scale features such as circulation, currents and river outflow, measurement of water quality, ship routeing, environmental monitoring of coastal zones, hazard assessment, and the management of fishing fleets. Examples: *AATSR, AVHRR, EOSP, MSU-M, Vegetation, VTIR.*

4. Instruments whose primary purpose is surface and cloud-top temperature measurement, in which thermal infrared bands predominate over others. These also have swath widths ranging typically from 500 to 3000 km. Examples: *ATSR, SCMR, SR, UHF Radiometer, VIRS.*

5. Ocean-colour instruments, which are similar to the general-purpose imagers but have narrower spectral bands (typically 20 nm or less) in the visible and near infrared parts of the spectrum, and swath widths usually ranging from 1000 to 3000 km. (The *MOS* instruments are an exception in having much narrower swaths.) These instruments have sufficient spectral resolution to distinguish the oceanic chlorophyll signal and to determine aerosol concentrations, in addition to the usual imaging functions. Examples: *CZCS, GLI, MKS-M2-BS, MOS, OCTS, POLDER, SeaWiFS, SROM.*

Multispectral Optoelectronic Scanner See *MOS.*

Multispectral Radiometer See *VISSR.*

Multispectral Scanner See *MSS.*

Multi-variable remote sensing The widely-used principle that a remote sensing observation that measures several properties for each pixel (such as the reflectance in several wavebands, the microwave emissivity at several frequencies, or the radar backscatter at different polarisations) can provide significantly more information about the target material than would be available if only a single property were measured. Most remote sensing instruments, with the exception of spaceborne *radar altimeters* and *synthetic aperture radars*, currently provide multi-variable measurements. Multitemporal observations, where the same instrument is used to obtain repeated observations of the same location at different times, can provide an extra dimension to the data set.

MVIRI See *VISSR*.

MVISR (Multispectral Visible and Infrared Scanning Radiometer) Chinese optical/ infrared imaging radiometer, proposed for inclusion on forthcoming *FY-1* satellites. Wavebands: 0.43–0.48, 0.48–0.53, 0.53–0.58, 0.58–0.68, 0.84–0.89, 0.90–0.97, 1.58–1.64, 3.55–3.93, 10.3–11.3, 11.5–12.5 μm. Spatial resolution: 1.1 km at nadir. Swath width: 3200 km.

The MVISR is an advanced version of the *VHRSR* carried on FY-1A and FY-1B. The function of the instrument is similar to the *AVHRR* and, as with the AVHRR, data can be downloaded in HRPT and APT formats.

MVZA Russian conically scanned passive microwave radiometer for sounding atmospheric water vapour content, planned for inclusion on *Meteor-3M* satellites.

Frequency (GHz)	Bandwidth (GHz)	Spatial resolution (km)
90.5	2.4	20
103.3 ± 1.0*	1.0	10
103.3 ± 4.0*	1.0	10
103.3 ± 8.0*	2.0	10
160.2	2.4	12

*Tuneable bands

Incidence angle: 42°. Swath width: 1600 km.

MWR (Microwave Radiometer) European nadir-viewing passive microwave radiometer, for *SST* and atmospheric water vapour estimation, planned for inclusion on *Envisat*. Frequencies: 23.8 and 36.5 GHz. Spatial resolution: 20 km.

URL: http://envisat.estec.esa.nl/instruments/mwr/index.html

MZOAS Russian conically scanned passive microwave radiometer planned for inclusion on *Meteor-3M* satellites.

Frequency (GHz)	Bandwidth (GHz)	Polarisation
5.0	0.1	H, V
10.7	0.1	H, V
10.7	0.2	H, V
22.2	0.4	H
36.5	1.0	H, V
90.5	2.4	H, V

Incidence angle: 42°. Swath width: 1500 km.

The MZOAS will be used for surface observations and for atmospheric sounding.

Nadir Vertically downwards, i.e. towards the centre of the Earth. The opposite direction to the nadir is the *zenith*.

NASA (National Aeronautics and Space Administration) The U.S. space agency, founded in 1958 from the National Advisory Committee for Aeronautics. Address: NASA Headquarters, Washington DC 20546, U.S.A.

URL: http://www.nasa.gov/

NASA Scatterometer See *NSCAT*.

NASDA The Japanese Space Agency, founded in 1969.
Address: National Space Development Agency of Japan (NASDA),
World Trade Center Bldg., 2-4-1,
Hamamatsu-cho, Minato-ku, Tokyo 105-60

URL: http://www.nasda.go.jp

National Space Agency of Ukraine See *CEOS*.

NAVSTAR See *GPS*.

NDVI Normalised difference vegetation index. See *vegetation index*.

Nearest-neighbour resampling See *resampling*.

Near infrared See *infrared*.

NEMS (Nimbus-E Microwave Spectrometer) U.S. nadir-viewing passive microwave radiometer, for profiling atmospheric temperature and water vapour and for surface observations, carried on *Nimbus-5*. Frequencies: 22.2, 31.4, 53.7, 54.9, 58.8 GHz. Spatial resolution: 180 km. Temperature resolution: 0.1–0.3 K. Absolute accuracy: 2 K.

Neva See *infrared radiometers*.

Nimbus-E Microwave Spectrometer See *NEMS*.

Nimbus-1 U.S. satellite, operated by *NASA*, launched in August 1964. Nimbus-1 provided useful data for about 1 month. Objectives: Experimental meteorological satellite. Orbit: Elliptical *LEO* at 490 to 1110 km altitude. Period 98 minutes; inclination 99°. Principal instruments: *APT, AVCS, HRIR*.

Nimbus-2 to -6 U.S. satellite series, operated by *NASA* and launched between May 1966 and June 1975. The satellites had lifetimes of between 2 and 10 years. Objectives: Experimental-operational meteorological satellites. Orbit: Nominally circular *LEO* (eccentricity close to zero for Nimbus-5 and -6) at minimum altitudes between 1100 and 1250 km, maximum altitudes between 1100 and 1350 km. Period 107 to 108 minutes; inclination 100° to 101°. Principal instruments: *AVCS* (Nimbus-2), *BUV* (Nimbus-4), *ERB* (Nimbus-6), *ESMR* (Nimbus-5 and -6), *HIRS* (Nimbus-6), *HRIR* (Nimbus-2), *IDCS* (Nimbus-3 and -4), *IRIS* (Nimbus-3 and -4), *ITPR* (Nimbus-5), *LRIR* (Nimbus-6), *MRIR* (Nimbus-2 and -3), *NEMS* (Nimbus-5), *PMR* (Nimbus-6), *SCAMS* (Nimbus-6), *SCMR* (Nimbus-5), *SCR* (Nimbus-4), *SIRS-A* (Nimbus-3), *SIRS-B* (Nimbus-4), *THIR* (Nimbus-4 and -5). Nimbus-3 onwards carried data collection packages.

Nimbus-7 U.S. satellite, operated by *NASA*, launched October 1978 and still partially functioning. Objectives: Operational meteorological and environmental observations (atmospheric trace gases, ocean colour, SST, sea ice etc.). Orbit: Circular *Sun-synchronous LEO* at 955 km altitude. Period 104.2 minutes; inclination 99.3°. Principal instruments: *CZCS* (functional to 1986), *ERB, LIMS, SAM II, SAMS* (functional to 1985), *SBUV, SMMR* (functional to 1988), *THIR* (functional to 1987), *TOMS*.

Nit The unit of luminance, equal to one *candela* per square metre. See *photometric quantities*.

Nitrous oxide A minor constituent of the Earth's *atmosphere*. Chemical symbol N_2O. Nitrous oxide can be detected in the infrared spectrum using broad absorption lines at 3.8 μm and 7.8 μm. It also has a microwave absorption line at 301 GHz.

NOAA (U.S.) National Oceanic and Atmospheric Administration, part of the US Department of Commerce.

URL: http://www.noaa.gov/

NOAA-1 to -5 U.S. satellite series operated by *NOAA*. The satellite lifetimes were typically 2 years. NOAA-1 was launched in December 1970 and NOAA-5 in July 1976. Objectives: Operational meteorological satellites. Orbit: Circular *LEO* at approximately 1500 km altitude. Period 115 to 116 minutes; inclination 99° to 102°. Principal instruments: *APT* (NOAA-1), *AVCS* (NOAA-1), *FPR* (NOAA-1), *SR, VHRR* (NOAA-2 to -5), *VTPR* (NOAA-2 to -5).

NOAA-6 to -8 U.S. satellite series operated by *NOAA*. The satellite lifetimes were typically 2 years. NOAA-5 was launched in July 1979 and NOAA-8 in March 1983. Objectives: Operational meteorological satellites. Orbit: Circular *Sun-synchronous LEO* at 800 to 850 km altitude. Period 101.2 to 101.9 minutes; inclination 98.2° to 98.9°. Principal instruments: *AVHRR, HIRS-2, MSU, SSU*. The satellites carried the *Argos* data collection system, and NOAA-8 also had a search and rescue package.

NOAA-8 onwards are designated ATN (Advanced TIROS-N) satellites.

NOAA-9 onwards U.S. satellite series operated by *NOAA*. Launch dates: NOAA-9, December 1984; NOAA-10, September 1987; NOAA-11, September 1988; NOAA-12, May 1991; NOAA-13, August 1993 (failed shortly after launch); NOAA-14, January 1995. NOAA-K was launched in May 1998 and will be redesignated NOAA-15 if successfully deployed. Operation of the series is scheduled at least until 2007. Objectives: Operational meteorological satellites. Orbit: Circular *Sun-synchronous LEO* at 858 km altitude. Period 102.2 minutes; inclination 98.9°; equator crossing time 13:50 ascending node (odd-numbered satellites), 07:30 descending node (even-numbered satellites). Principal instruments: *AVHRR, ERBE, SBUV, TOVS*. The satellites also carry *Argos*, solar–terrestrial physics and search and rescue packages.

The fact that two satellites are in operation at any one time means that at least two overpasses per day (morning and afternoon) are possible for any point within the latitudinal limits. These NOAA satellites are also called POES (Polar-orbiting Operational Environmental Satellites), and have evolved from the earlier *TIROS* and *ESSA* series and the *ITOS* satellite.

URL: http://psbsgi1.nesdis.noaa.gov:8080/EBB/ml/nic00.html
http://www.itc.nl/~bakker/noaa.html

Nodal period The time between successive passages of a satellite through its *ascending node* (or, equivalently, its *descending node*). It is given with sufficient accuracy for most purposes by

$$P_{\mathrm{n}} = 2\pi\sqrt{\frac{a^3}{GM}\left(1 + \frac{3J_2 a_{\mathrm{e}}^2}{4a^2}\left(1 - 3\cos^2 i + \frac{1 - 5\cos^2 i}{(1 - e^2)^2}\right)\right)},$$

where a is the semimajor axis, e the eccentricity and i the *inclination* of the orbit. J_2 is the Earth's *dynamical form factor*, a_{e} is its equatorial radius, M is its mass, and G is the gravitational constant. For circular orbits ($e = 0$) the formula can be written numerically as

$$P_{\mathrm{n}} = 0.009\,952 a^{3/2}\left(1 + \frac{66\,063(1 - 4\cos^2 i)}{a^2}\right),$$

where P_{n} is measured in seconds and a is measured in kilometres.

Nodes The nodes of a satellite's *orbit* are the points where it intersects the Earth's equatorial plane. See *ascending node, descending node*.

138

Nonesuch See *tasselled-cap transformation.*

Non-imaging sensors Remote sensing instruments that do not generate an *image*, either because they do not provide a two-dimensional representation of the Earth's surface or because the intensity of the detected radiation is not the main parameter of interest. The main types of non-imaging sensor are *atmospheric sounders, radar altimeters, scatterometers* and nadir-viewing radiometers.

Non-linear contrast enhancement See *contrast enhancement.*

Normalised difference vegetation index See *vegetation index.*

NSAU (National Space Agency of Ukraine) See *CEOS.*

NSCAT (NASA Scatterometer) U.S. microwave wind scatterometer, carried on *ADEOS* satellite. Frequency: K_u band (14.0 GHz). Beam angles: [$0° =$ forward, $90° =$ right] $±45°$, $−65°$, $+115°$, $±135°$. Spatial resolution: 50 km (data provided on 25 km grid). Swath width: 600 km. Near edges 150 km from sub-satellite track; far edges 750 km. Left and right. Accuracy: $±2$ m/s (3–20 m/s), $±10\%$ (20–30 m/s); $±20°$ (3–30 m/s) for ocean surface wind vectors.

NSCAT is based on the *Seasat SASS* instrument. A modified version, originally called NSCAT-II, is now called *SeaWinds.*

URL: http://hdsn.eoc.nasda.go.jp/guide/guide/satellite/sendata/nscat_e.html
http://winds.jpl.nasa.gov/

Nyquist frequency The minimum frequency at which a signal must be sampled if it is to be possible to reconstruct the signal from the samples. If the signal is band-limited such that it contains no frequency components (see *Fourier transform*) above f_{max}, the Nyquist frequency is $2f_{max}$. See also *aliasing.*

Occultation In the context of Earth remote sensing, occultation occurs when the line of sight from a celestial object (Sun, Moon or star) to an instrument is intercepted by the Earth. When the celestial object is almost, but not completely, occulted, the line of sight through the atmosphere is very long, and consequently the detected radiation contains information about the atmosphere. This is the basis of atmospheric *limb-sounding*.

Ocean colour Ocean colour measurements are narrow-band multispectral reflectance observations of the ocean surface in the visible and near infrared regions of the electromagnetic spectrum (see *multispectral imager*). The principal use of ocean colour measurements, initially developed using the *CZCS* instrument, is to estimate *biological productivity* in oceans by detecting the presence of photosynthetically active pigments, mainly chlorophyll-a (which has an absorption maximum at 443 nm) and phaeophytin. This requires very accurate *atmospheric correction*, because the atmosphere contributes roughly 90% of the radiance detected at a satellite. Ocean colour measurements can also be used in coastal zones to monitor the distribution of suspended sediments, and to observe marine pollution.

In ocean colour analysis it is usual to distinguish **Case 1 waters**, in which the surface reflectance is dominated by photosynthetically active pigments, from **Case 2 waters**, in which the reflectance is mostly contributed by suspended sediments and dissolved organic matter (**gelbstoff**). Case 2 waters are usually, though not always, coastal. Retrieval algorithms for case 1 waters generally assume that the surface concentration of chlorophyll is given by some empirical function of the upwelling radiances at the water surface, measured at two or three spectral bands. These are normally bands near 443 and 550 nm, though a band near 520 nm is also used.

The first spaceborne instrument optimised for ocean colour observations was CZCS (1978–1986). The Russian space station *Mir* (1986 onwards) carries two types of ocean colour instrument, *MKS-M2-BS* and *MOS*, and the short-lived *ADEOS-I* satellite carried the *OCTS* instrument. Continuity was maintained with the deployment of *SeaWiFS* in 1997, and future ocean colour instruments include *GLI, MERIS, MODIS, POLDER* and *SROM*.

Ocean Colour and Temperature Scanner See *OCTS*.

Ocean currents and fronts Boundaries between different water masses can be detected in remote sensing imagery by a variety of mechanisms. The water masses will often have different temperatures, detectable in thermal infrared imagery, or different quantities and types of suspended sediments, detectable in visible-wavelength imagery, especially *ocean colour* images. The boundaries of ocean currents can sometimes be detected in synthetic aperture radar images as a discontinuity in the radar backscatter, due for example to a change in the surface roughness as a result of the different wind velocity relative to the surface velocity on either side of the boundary.

The velocity of an ocean current other than at the equator can be estimated from *radar altimeter* measurements of the *sea surface topography*, using the reasonably accurate assumption that the surface is in geostrophic equilibrium. In a Cartesian coordinate system in which the x-axis points east, the y-axis north and the z-axis measures the height of the surface above the *geoid*, the components v_x and v_y of the surface velocity are given by

$$\frac{\partial z}{\partial x} = \frac{f}{g} v_y,$$

$$\frac{\partial z}{\partial y} = -\frac{f}{g} v_x,$$

where g is the gravitational field strength and f is the **Coriolis parameter**, defined as

$$f = 2\Omega \sin \lambda,$$

where Ω is the Earth's angular velocity of rotation and λ is the latitude. High-precision altimetric measurements are needed to apply this method to the determination of ocean currents.

Knowledge of ocean circulation, which on average transports energy polewards from the tropics, is important in studies of the global climate system. It also finds applications to ship routeing, off-shore exploration, and the routeing of sea-floor pipelines.

Ocean waves See *wave height and wave spectra*.

OCTS (Ocean Colour and Temperature Scanner) Japanese optical/infrared mechanically scanned imaging radiometer, carried on *ADEOS* satellite. Wavebands: 402–422, 433–453, 480–500, 510–530, 555–575, 655–675, 745–785, 845–885 nm; 3.55–3.88, 8.25–8.80, 10.3–11.4, 11.4–12.5 μm. Spatial resolution: 700 m. Swath width: 1400 km.

The sensor could be steered up to 20° fore and aft to avoid Sun glint. Temperature sensitivity (thermal infrared) 0.15 K. Optical/near infrared channels intended primarily for ocean colour measurements.

URL: http://hdsn.eoc.nasda.go.jp/guide/guide/satellite/sendata/octs_e.html

ODIN Swedish satellite planned for launch in 1998. Objectives: atmospheric sounding. Orbit: circular *Sun-synchronous LEO* at 600 km altitude. Period 97

minutes; inclination 97.7°; equator-crossing time 18:00 (ascending node). Principal instruments: *OSIRIS*, *SMR*.

ODUS (Ozone Dynamics Ultraviolet Spectrometer) Japanese nadir-viewing (backscatter) ultraviolet spectrometer, planned for inclusion on *EOS-Chem-1* satellite. ODUS, which is similar in concept to *BUV*, *SBUV* and *TOMS*, is intended to provide total (integrated) atmospheric ozone, NO_2 and SO_2 concentrations. Wavebands: 10 channels (0.5 nm resolution) between 306 and 410 nm. Spatial resolution: 20 km.

OEA (Optical Electronic Apparatus) Russian optical/near infrared *CCD* imaging radiometer on *Almaz-1B* satellite. Wavebands: 0.5–0.6, 0.6–0.7, 0.7–0.8, 0.58–0.8 μm. Spatial resolution: 2.5 m. Swath width: 80 km, selectable within ±300 km of the sub-satellite track by rolling the spacecraft.

There are two imagers, looking 25° forward and aft of nadir, to give a stereo viewing capability. The on-board tape recorder can store up to 12 minutes of data per orbit for downlinking. The instrument was formerly known as OSSI.

Oil spills Oil spills on water are classified as **slicks** if they are thick enough to present a black or brown visual appearance, **sheens** if they appear silvery, and **rainbows** if they are thin enough to show interference colours. 90% of the volume of a typical marine spill is in the form of a slick occupying 10% of the total area.

Oil spills can be detected by remote sensing at a range of wavelengths. Sheens of 0.15 μm or thicker can be detected at ultraviolet or near infrared wavelengths by virtue of their higher reflectance than open water: there is a somewhat weaker effect in blue light. Thermal infrared radiometry can reveal the presence of oil films more than about 10 μm thick, since the *emissivity* of an oil film (≈0.97) is less than that of open water (≈0.99), but with the possibility of confusion with cold water currents. Imaging *radar* can provide some of the clearest evidence for the presence of oil films of 1 μm thickness or greater, except under flat calm conditions. The presence of the film suppresses the short surface waves, increasing the specular backscatter away from the radar and consequently decreasing the signal detected by the radar. *Passive microwave radiometry* also has the possibility to detect oil films, although the mechanism is more complicated since the oil surface has a higher emissivity than open water, but also has a slight cooling effect on a rough water surface. Furthermore, the spatial resolution of passive microwave systems is generally poor compared with other remote sensing systems, especially when operated from satellites.

Okean-O Russian satellite programme scheduled to begin operation as soon as funds are made available. Objectives: Operational oceanographic and sea ice monitoring (development of *Okean-O1* programme). Orbit: Circular *LEO* at 650 km altitude. Period 98 minutes; inclination 82.5°. Principal instruments: *DELTA-2*, *MSU-M*, *MSU-SK*, *MSU-V*, *R225*, *R600*, *RLSBO*. The satellites will also carry a data collection package and an experimental visible-wavelength polarimeter (Trasser).

Okean-O1 Russian/Ukrainian satellite programme, begun (with launch of Okean-O1-1) in July 1988. The satellites have a nominal lifetime of two years. Objectives: Operational oceanographic and sea ice monitoring. Orbit: Circular *LEO* at 650 km altitude. Period 98 minutes; inclination 82.5°. Principal instruments: *MSU-M, MSU-S, RLSBO, RM-08*. The satellites also carry data collection packages.

Image data are downlinked at 137 MHz and freely accessible.

OLS (Operational Linescan System) U.S. optical/infrared mechanically scanned imaging radiometer, carried on *DMSP* Block 5D satellites. Wavebands: 0.4–1.1 μm, 8–13 μm (10.5–12.6 from June 1979, 10.2–12.8 from June 1987). Block 5D-2 satellites also have a 0.5–0.9 μm band with sufficient sensitivity to detect reflected moonlight. Spatial resolution: 0.6 km (daytime), 2.8 km (night time imagery stored on board). Swath width: 3000 km.

OMI (Ozone Monitoring Instrument) European nadir-looking ultraviolet/optical/near infrared spectrometer, planned for inclusion on *Metop-1*. Wavebands: 240 to 790 nm. Spectral resolution: 0.2 to 0.4 nm. Spatial resolution: 40 km × 40 km to 40 km × 320 km. Height resolution: 1 km.

OMI is based on the design of the *GOME* instrument.

Operational Linescan System See *OLS*.

OPS (Optical Sensor) Japanese optical/infrared imaging radiometer carried by *JERS-1* Satellite. Wavebands: 1: 0.52–0.60 μm; 2: 0.63–0.69 μm; 3 and 4: 0.76–0.86 μm; 5: 1.60–1.71 μm; 6: 2.01–2.12 μm; 7: 2.13–2.25 μm; 8: 2.27–2.40 μm. Spatial resolution: 18 × 24 m. Swath width: 75 km.

The band-4 sensor looks 50° to 75° forward from nadir, to give stereo viewing when combined with band 3.

Optical depth The optical depth of a point in the atmosphere is defined as the *optical thickness* of the atmosphere between that point and the top of the atmosphere.

Optical Electronic Apparatus See *OEA*.

Optical radiation See *visible-wavelength radiation*.

Optical Sensor See *OPS*.

Optical thickness The optical thickness τ of a medium is defined as

$$\tau = \int \gamma(x)\, dx,$$

where $\gamma(x)$ is the *attenuation* coefficient at position x and the integration is carried out over the entire medium. By replacing γ by the absorption coefficient γ_a or the scattering coefficient γ_s, optical thicknesses for absorption and scattering can also be defined.

See also *Lambert–Bouguer law*, *radiative transfer equation*.

Orbimage Commercial satellite imagery company, based in Dulles, Virginia, U.S.A.

URL: http://www.orbimage.com

Orbital manoeuvres Adjustments to the parameters of a satellite *orbit*, for example to correct it after initial injection by the launch vehicle or for the effects of perturbing influences. Satellites usually carry small propulsion systems to allow these manoeuvres to take place. The amount of fuel required is determined by the total velocity increment for the manoeuvre, equal to the sum of the magnitudes of all the changes in velocity required to perform it. The table below gives the velocity increments Δv necessary for small changes in the radius a, period P, eccentricity e and inclination i of a nearly circular orbit with orbital velocity v.

Parameter	Δv
P	$\dfrac{v}{3}\dfrac{\Delta P}{P}$
a	$\dfrac{v}{2}\dfrac{\Delta a}{a}$
e	$\dfrac{v}{2}\Delta e$
i	$2v\sin\dfrac{\Delta i}{2}$

The mass of fuel required to achieve a velocity increment Δv is given by

$$\frac{M\Delta v}{gI},$$

where M is the satellite mass, g is the acceleration due to gravity at the Earth's surface and I is the **specific impulse** of the propellant. Typical propellants have specific impulses between 200 and 300 seconds.

Orbit, satellite The curved path described by a satellite moving round the Earth, principally under the influence of gravitational forces. To a very good approximation, the orbit is elliptical, so that its shape and size are specified by the following parameters: **semimajor axis**; **semiminor axis**; **eccentricity**. These are illustrated in figure A.

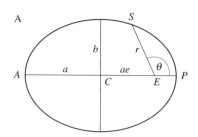

E represents the position of the Earth's centre of mass, located at one focus of the ellipse. P and A are respectively the *perigee* and *apogee* of the orbit. C is the centre of the ellipse, and a and b are respectively the semimajor and semiminor axes. a, b and the eccentricity e are related by

$$b^2 = a^2(1 - e^2).$$

For a circular orbit, $a = b$ and $e = 0$.

A general position S of the satellite is represented by the polar coordinates (r, θ), where

$$r = \frac{a(1 - e^2)}{1 + e \cos \theta}.$$

This formula specifies the shape of the orbit, but not the rate at which the satellite travels along it. The satellite's position θ depends on the time t according to the following formula:

$$\frac{t}{P_n} = \frac{1}{\pi} \arctan \frac{(1 - e) \tan(\theta/2)}{\sqrt{1 - e^2}} - \frac{e}{2\pi} \frac{\sqrt{1 - e^2} \sin \theta}{1 + e \cos \theta},$$

where P_n is the satellite's *nodal period*. For small e, this can be expanded to give the following formula for θ in terms of t:

$$\theta = \frac{2\pi t}{P_n} + 2e \sin \left(\frac{2\pi t}{P_n} \right) + \frac{5e^2}{4} \sin \left(\frac{4\pi t}{P_n} \right) + \cdots.$$

This formula gives a maximum error of approximately $4e^3/3$ radians.

The position and orientation of the orbit with respect to the Earth are specified by its *inclination*, the longitude of the *ascending node*, and the position of the perigee. A circular orbit has no perigee, so it is specified only by its inclination and by the longitude of the ascending node.

Circular orbits. A circular orbit is specified by its radius a (or equivalently by the orbital height, conventionally defined as $a - a_e$, where a_e is the Earth's equatorial radius), its inclination, and the geographical coordinates of the *sub-satellite point* at some specified date and time. Subsequent variation of the coordinates of the sub-satellite point is determined by: (i) motion of the satellite in its orbit; (ii) rotation of the Earth; and (iii) *precession* of the orbital plane. Effects (ii) and (iii) can be combined to give the longitudinal rate at which the ascending node of the orbit moves relative to the Earth's surface; this is given by $\Omega_e - \Omega$ in a westward sense, where Ω_e is the Earth's rotational angular velocity of 2π radians per *sidereal day*, and Ω is the angular velocity of the precession. Effect (i) can be calculated by spherical trigonometry, as shown in figure B.

In this diagram, N represents the North Pole, E the Earth's centre, A the instantaneous position of the ascending node and S the satellite's instantaneous position. The inclination of the orbit is represented by i, and ϕ is the angle subtended at E between S and A. To a good approximation, ϕ can be calculated from the time since the satellite passed through its ascending node by assuming that ϕ varies uniformly with time, increasing by 2π radians in each nodal

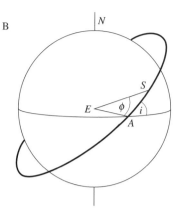

B

N

E

S

φ

i

A

period. The latitude B of the sub-satellite point is then given by

$$\sin B = \sin \phi \sin i$$

(where positive B corresponds to the Northern hemisphere). If ϕ is expressed in the range $0 \le \phi < 2\pi$, the longitude L of the sub-satellite point is given by

$$L_A + \arccos \frac{\cos \phi}{\cos B}$$

if $\phi \le \pi$ and $i \le \pi/2$ or if $\phi > \pi$ and $i > \pi/2$, and by

$$L_A + 2\pi - \arccos \frac{\cos \phi}{\cos B}$$

if $\phi > \pi$ and $i > \pi/2$. In these expressions, longitude is defined as positive to the east, and L_A is the longitude of the ascending node A.

Careful choice of the radius and inclination of a circular satellite orbit allow orbits with particularly useful properties to be obtained. Particularly important amongst these special orbits are *geosynchronous*, *Sun-synchronous* and *exactly repeating* orbits.

Orbview 3 U.S. satellite planned for launch in 1998, operated by *Orbimage*. Objectives: High resolution Earth observation for commercial use. Orbit: Circular *Sun-synchronous LEO* at 470 km altitude. Equator crossing time 10:30 a.m.

Orbview 3 will carry a high resolution optical/near infrared imager operating in panchromatic and multispectral modes. Wavebands: 0.45–0.90 µm (panchromatic); 0.45–0.52, 0.52–0.60, 0.63–0.69, 0.76–0.90 µm (multispectral). Spatial resolution: 1 m (panchromatic), 4 m (multispectral). Swath width: 8 km.

OSIRIS Swedish ultraviolet/optical/infrared limb sounder, planned for inclusion on the *ODIN* satellite. Wavebands: 0.28–0.45, 0.45–0.80, 1.27 µm. Spectral resolution: 1 nm, 2 nm, 10 nm respectively. The instrument will be used to derive atmospheric concentration profiles for oxygen, ozone, nitrous oxide and other trace gases.

OSSI See *OEA*.

Outer Space Treaty See *Legal and international aspects*.

Oxygen A major constituent of the Earth's *atmosphere*. Molecular formula O_2. Oxygen is well mixed in the atmosphere, so that its density at a point is proportional to the atmospheric density. The principal atmospheric absorption lines that are used for the detection and measurement of oxygen are at wavelengths of 0.69, 0.76 and 1.25 µm in the visible and infrared part of the spectrum, and at 60, 119 and 298 GHz in the microwave region.

Ozone An important minor constituent of the Earth's *atmosphere*. Molecular formula O_3. Ozone occurs naturally in the *stratosphere* as a result of the action of solar ultraviolet and X-ray radiation on oxygen. This ozone layer effectively removes such radiation from the component reaching the Earth's surface. Ozone also occurs in the *troposphere*. In the upper troposphere and lower stratosphere, ozone is an effective 'greenhouse gas'. *Atmospheric sounding* of ozone is important for understanding the Earth's radiation budget and stratospheric *chemistry*. Atmospheric ozone concentrations can be measured using both absorption and scattering techniques. The main absorption lines used for ozone sounding are at 0.26 (Hartley bands), 0.65 (Chappuis band), 4.7, 9.6 and 14.1 µm in the ultraviolet, visible and infrared region of the electromagnetic spectrum. In the microwave region, the main absorption lines are at 102, 142, 166, 184, 195, 243, 248, 289 and 302 GHz.

Ozone Dynamics Ultraviolet Explorer See *ODUS*.

Ozone Monitoring Instrument See *OMI*.

Ozon-M Russian ultraviolet/optical/near infrared Sun-viewing limb sounder, for measurement of atmospheric ozone and other gases, carried on *Mir-1*. Wavebands: 0.25–0.29, 0.37–0.39, 0.60–0.64, 0.99–1.03 µm (total 188 channels). Spectral resolution: 0.3 to 1.0 nm. Spatial resolution: 15 km (horizontal), 1 km (vertical). Height range: 5–70 km.

PAN (Panchromatic camera) Indian optical/near infrared single channel *CCD* (pushbroom) imaging radiometer, carried on *IRS-1C* and *-1D*. Waveband: 0.50–0.75 µm. Spatial resolution: 5 m. Swath width: 71 km, steerable ±26° from sub-satellite track.

Panchromatic Term originating in photography, usually meaning a broad range of spectral sensitivity roughly corresponding to that of the human eye (about 0.4 to 0.65 µm). See *photometric quantities*.

Panchromatic camera See *PAN*.

PAR See *photosynthetically active radiation*.

Parallelepiped classifier See *supervised classification*.

Parallel polarisation Radiation that is incident on the planar interface between two media is said to be parallel polarised if the electric field vector of the radiation is parallel to the plane containing the surface normal and the wave vector of the incident radiation. If the interface is approximately horizontal, the radiation is also said to be vertically polarised.

Passive microwave radiometry A remote sensing technique that uses microwave and millimetre-wave radiometers to measure the *black-body radiation* emitted by natural or man-made objects. At frequencies below 300 GHz and for objects with physical temperatures greater than 200 K, the *Planck* radiation law reduces to the approximate form given by the *Rayleigh–Jeans* formula, which exhibits a linear relationship between the *radiance* of a black body and its physical temperature T. For a real body with an *emissivity* ε, this linear relationship leads to the definition of the *brightness temperature* $T_b = \varepsilon T$, as representative of the radiance emitted by the body. (See *antenna temperature*.)

Microwave radiometers have been flown on numerous atmospheric and oceanographic satellites for measuring the temperature profile of the *atmosphere*, the integrated *water vapour* content, *cloud* water content, *sea surface temperature*, *sea ice* type and distribution etc.

The following table summarises the characteristics of the principal spaceborne passive microwave radiometers. Instruments for which the spatial

Instrument	Launch (year)	Frequencies (GHz) and polarisations	Spatial resolution (km)	Swath width (km)	Applications
ESMR	1972	19.4 H	25	3000	Surface
NEMS	1972	22.2, 31.4, 53.7, 54.9, 58.8	180	180	Surface, atmosphere
SHF	1974	7.5, 19.4, 22.2, 37.5	13–30	800	Surface, atmosphere
ESMR	1975	37 H, V	20	1300	Surface
SCAMS	1975	22.2, 31.7, 52.9, 53.9, 55.5	145	2600	Surface, atmosphere
MSU	1978	50.3, 53.7, 55.0, 57.9	100	2400	Atmosphere
SMMR	1978	6.6, 10.7, 18, 21, 37 H, V	30–140	600–800	Surface, atmosphere
SSM/T	1978	50.5, 53.2, 54.3, 54.9, 58.4, 58.8, 59.4	175	1500	Atmosphere
MSR	1987	23.8, 31.4	23–32	320	Surface, atmosphere
SSM/I	1987	19.4, 22.2, 37.0, 85.5 H, V	16–70	1400	Surface, atmosphere
RM-08	1988	36.6	20	550	Surface
ATSR-MWS	1991	23.8, 36.5	20	20	Atmosphere
MLS	1991	63, 183, 205	Atmospheric limb sounder		Surface
UHF radiometer	1991	6.0, 37.5	10–30	500	Atmosphere
TMR	1992	18, 21, 37	30	30	Surface, atmosphere
IKAR-Delta	1995	7.5, 22.2, 37.5, 100	5–50	400	Surface, atmosphere
IKAR-N	1995	5.0, 13.3, 22.2, 37.5, 100	5–75	60–750	Surface
IKAR-P	1995	5, 13.3 H, V	75	750	Surface
R400	1995	7.5	50	400	Surface
SMR	1997	119, 486–504, 541–586	Atmospheric limb sounder		Atmosphere (rain)
TMI	1997	10.7, 19.4, 22, 37, 88 H, V	4–45	680	Atmosphere
AMAS	1999	298–626	Atmospheric limb sounder		Atmosphere
AMR	1999	18.2, 23.8, 34.0	16–29	16–29	Surface, atmosphere
AMSR	1999	6.6, 10.7, 18.7, 23.8, 36.5, 55, 89	5–60	1700	Surface, atmosphere
MIVZA	1999	20, 35, 94 H, V	50–100	1350	Atmosphere
MTZA	1999	18.7, 36.5, 52.2, 55.4, 57.0, 90.5	20–90	1500	Atmosphere (water)
MVZA	1999	90.5, 103.3, 160.2	10–20	1600	Atmosphere
AMSU	2000	23.8–31.4 (15 bands)	40	2200	Atmosphere
MHS	2000	89, 166, 183	14	1650	Atmosphere
MIMR	2000	6.8, 10.7, 18.7, 23.8, 36.5, 89 H, V	5–60	1400	Surface, atmosphere
MWR	2000	23.8, 36.5	20	20	Atmosphere

resolution and swath width are identical view the nadir only: others are downward-looking scanning instruments unless otherwise indicated.

Passive system A remote sensing system that detects naturally occurring radiation, either reflected sunlight (optical and near infrared systems) or thermal radiation (thermal infrared and passive microwave systems). Compare *active system*.

P band Subdivision of the microwave region of the *electromagnetic spectrum*, covering the frequency range 225 to 390 MHz (wavelengths 0.77 to 1.33 m).

Perigee The closest point of a satellite's *orbit* to the Earth's centre.

Period See *orbit, satellite*.

Permeability, magnetic The magnetic permeability μ of a material is the ratio of the magnetic flux density B to the magnetic field strength H inducing it. It can be written as

$$\mu = \mu_0 \mu_r,$$

where μ_0 is the magnetic permeability of free space and μ_r is the (dimensionless) relative magnetic permeability. μ_r is less than 1 for diamagnetic materials, greater than 1 for paramagnetic materials and much greater than 1, and dependent on the magnetic field strength, for ferromagnetic materials.

Permittivity, electric See *dielectric constant*.

Perpendicular polarisation Radiation that is incident on the planar interface between two media is said to be perpendicularly polarised if the electric field vector of the radiation is perpendicular to the plane containing the surface normal and the wave vector of the incident radiation (i.e. parallel to the plane of the interface). If the interface is approximately horizontal, the radiation is also said to be horizontally polarised.

Perpendicular vegetation index See *vegetation index*.

Pests and diseases, agricultural Remote sensing methods have proved valuable in detecting agricultural areas affected by pests (especially insect pests) and diseases. Detection methods include characteristic changes in the optical and near infrared reflectance of the vegetation, defoliation, and changes in the geometry of the crowns of trees.

Phase Russian optical/near infrared spectrometer, for atmospheric profiling and surface characterisation, carried on *Mir-1*. Waveband: 0.445–2.2 µm. Spectral resolution: 15 nm. Spatial resolution: 1 km.

Phase function See *radiative transfer equation.*

Phase velocity See *dispersion.*

Phoenix See *infrared radiometers.*

Photodiode A type of detector used in some *electro-optical* systems. The photo-diode is a semiconductor device, which can operate either in photovoltaic mode, in which the voltage appearing across the terminals of the device is proportional to the intensity of the radiation incident upon it, or in photocon-ductive mode, in which the incident radiation generates an intensity-dependent current through the device. Photodiode detectors have the advantage of very small size, robustness, rapid response, and sensitivity to wavelengths up to about 5 μm.

Photogrammetry The technique of obtaining quantitative measurements of the geometric properties of a target area from one or more photographs of the area. The relationship between the geometrical properties of the target and those of its photographic image is simple and well defined, unlike the relationship between corresponding radiometric properties, so that aerial or space photo-graphs are well suited to planimetric mapping applications.

A vertical aerial photograph, i.e. one in which the camera's optical axis is precisely perpendicular to a horizontal plane, has a *scale* that is determined only by the camera's focal length f and its height H above the horizontal plane. The scale of the negative, or of a contact print made from the negative, is

$$\frac{f}{H},$$

so if the target surface is confined to the horizontal plane there is a fixed one-to-one correspondence (fixed scale) between the spatial coordinate on the ground and the corresponding spatial coordinate in the photograph. The scale can be calculated from f and H if they are known with sufficient accuracy, or can be determined from *ground control points.*

If the target surface has relief, the relationship between target coordinates and photograph coordinates is more complicated. A point located at (x, y, h), where x and y are the Cartesian coordinates measured in two orthogonal horizontal directions with respect to the nadir point (the point on the surface directly below the camera, assumed to be coincident with the **principal point**), and h is the height measured above the horizontal plane, will be imaged in the negative at coordinates

$$\left(\frac{fx}{H-h}, \frac{fy}{H-h} \right),$$

provided that the line of sight from the point to the camera is not obscured by other parts of the surface. The displacement of this point from the coordinates $(fx/H, fy/H)$ that it would have if $h = 0$ is termed the *relief displacement,* and contains information about the height h. This is the basis of *stereophotography.*

The above formulae become considerably more complicated in the case of an oblique (i.e. non-vertical) aerial photograph.

Photographic infrared See *infrared*.

Photographic systems Photographic systems still find spaceborne (and, to a much greater extent, airborne) applications, despite the advantages (digital, calibrated output) offered by *electro-optical systems* operating in the same waveband (optical to near infrared). Although a photograph is not recorded in the form of pixels, its spatial resolution can be defined by specifying the number of pixels that would be needed in an electro-optically acquired image having the same spatial coverage and resolution. This number falls typically in the range 30 million to 300 million, compared with at most 40 million for most electro-optical systems. In addition, photographic systems offer high geometric accuracy, minimising the need for *geometric correction* and suiting them to mapping applications (including *stereophotography*), and comparatively low cost.

Since the photographic image is recorded physically rather than electronically, it cannot be transmitted to a receiving station by radio signal. For this reason, photographic systems are primarily used on recoverable and reusable platforms such as the *Space Shuttle* and the *Mir* space station. The following table summarises the main spaceborne photographic systems, excluding cameras used for hand-held photography by astronauts:

Satellite	Launch	Camera	Coverage (km)	Spatial resolution (m)
Space Shuttle	1981	LFC	170×340	15
Spacelab	1983	MC	189×189	12
Mir-1	1986	KAP-350	180×180	30
Mir-1	1986	MKF-6MA	145×210	30
Mir-1	1990	KFA-1000	105×105	20
Resurs-F	1991	KFA-200	245×245	30
Resurs-F	1991	MK-4	144×144	14

Photometric quantities Photometric quantities are analogous to the radiometric quantities (*irradiance, radiance, radiant exitance, radiant flux* and *radiant intensity*) but are weighted according to the nominal sensitivity of the human eye. The quantity equivalent to (radiant) power is the luminous flux ϕ_l, defined by

$$\phi_l = K \int_0^\infty \phi_\lambda V(\lambda)\, d\lambda,$$

where K is a constant with a value of 680 lumens per watt, ϕ_λ is the radiant power per unit wavelength interval, and $V(\lambda)$ is a function (shown below) of wavelength that represents the nominal sensitivity of the eye.

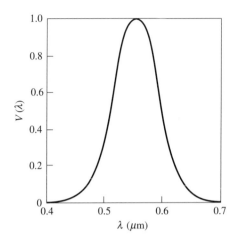

The photometric quantities analogous to radiometric quantities, together with their units, are given below.

Radiometric quantity	Photometric equivalent	Unit
power	luminous flux	lumen (lm)
radiant intensity	luminous intensity	candela (cd)
radiance	luminance	nit
irradiance	illuminance	lux (lx)
radiant exitance	luminous exitance	

Photometric quantities are sometimes used to describe the performance of systems operating in the visible part of the electromagnetic spectrum, particularly *panchromatic* photographic systems.

Photomultiplier A type of detector used in some *electro-optical* systems. The photomultiplier is a vacuum-tube device in which incident photons are detected by means of the photoelectric effect. A series of accelerating anodes amplifies the electron current to a detectable level that is proportional to the intensity of the incident radiation. Photomultipliers are very sensitive, have rapid response times, and can operate at wavelengths up to about 1 μm. Their main disadvantages compared with *photodiode* detectors is their larger size and relative lack of mechanical robustness.

Photosynthetically active radiation Photosynthetically active radiation (PAR) is light having a wavelength between about 0.4 and 0.7 μm. Since the amount of photosynthesis (number of photochemical reactions) in a plant is roughly proportional to the number of photons of PAR absorbed, a useful measure of PAR is *Einsteins* per square metre per second. Radiometers designed for measuring PAR should have a spectral response proportional to wavelength between 0.4 and 0.7 μm.

Physical optics model See *Kirchhoff model*.

Phytoplankton Chlorophyll-containing micro-organisms (algae) in the sea or other water body. Phytoplankton are the primary biological organisms on which other marine life depends, and are important in biogeochemical processes, particularly the fixing of dissolved carbon as carbonates. See *biological productivity, ocean colour*.

Pixel Pixels are the discrete elements ('pixel' = 'picture element') of which digital *images* are composed. Each pixel is specified by coordinates representing its position in the image (and hence on the ground), and by one or more values representing the values of the physical properties measured by the sensor. For example, a pixel of a *Landsat TM* image is specified by its position within the image and by seven digital numbers representing the *spectral radiance* of the detected radiation in each of the seven spectral bands of the instrument. The size of the region of the Earth's surface corresponding to a pixel is one definition of the spatial *resolution* of the instrument.

Planck distribution The *spectral radiance* of a *black body* at (absolute) temperature T is given by the Planck distribution

$$L_\nu = \frac{2h\nu^3}{c^2(\exp[h\nu/kT] - 1)}$$

or equivalently

$$L_\lambda = \frac{2hc^2}{\lambda^5(\exp[hc/\lambda kT] - 1)},$$

where ν is the frequency, λ is the wavelength, h is Planck's constant, c is the speed of light and k is Boltzmann's constant. See *Stefan's law, Wien's law, Rayleigh–Jeans approximation*.

The integral of the Planck distribution over all frequencies (or wavelengths) is

$$\int_0^\infty L_\nu \, d\nu = \frac{2\pi^4 k^4 T^4}{15c^2 h^3} = \frac{\sigma T^4}{\pi},$$

where σ is the Stefan–Boltzmann constant. The fraction $f(x)$ of this total radiation that is emitted between frequencies 0 and ν is given by the table below, in terms of the dimensionless parameter $x = h\nu/kT$. The tabulation at intervals $\Delta x = 0.1$ is sufficient to allow quadratic interpolation to five-figure accuracy.

For small values of x, $f(x)$ can be evaluated more accurately using the series expansion

$$f(x) = \frac{15}{\pi^4}\left(\frac{x^3}{3} - \frac{x^4}{8} + \frac{x^5}{60} - \frac{x^7}{5040} + \frac{x^9}{272\,160} - \frac{x^{11}}{13\,305\,600} + \frac{x^{13}}{622\,702\,080} \cdots\right),$$

and for large values of x,

$$1 - f(x) = \frac{15}{\pi^4} \sum_{j=1}^\infty \left(\frac{x^3}{j} + \frac{3x^2}{j^2} + \frac{6x}{j^3} + \frac{6}{j^4}\right)\exp(-jx).$$

	.0	.1	.2	.3	.4	.5	.6	.7	.8	.9
0	0.00000	0.00005	0.00009	0.00124	0.00282	0.00529	0.00879	0.01341	0.01923	0.02629
1	0.03462	0.04421	0.05506	0.06714	0.08040	0.09478	0.11023	0.12667	0.14402	0.16221
2	0.18115	0.20074	0.22092	0.24158	0.26264	0.28403	0.30565	0.32743	0.34930	0.37118
3	0.39302	0.41473	0.43628	0.45760	0.47865	0.49938	0.51975	0.53973	0.55929	0.57840
4	0.59703	0.61516	0.63279	0.64990	0.66647	0.68251	0.69800	0.71294	0.72735	0.74121
5	0.75453	0.76733	0.77960	0.79135	0.80260	0.81336	0.82363	0.83344	0.84278	0.85169
6	0.86016	0.86822	0.87588	0.88316	0.89006	0.89660	0.90280	0.90866	0.91422	0.91947
7	0.92443	0.92911	0.93353	0.93770	0.94164	0.94534	0.94883	0.95211	0.95520	0.95811
8	0.96084	0.96340	0.96581	0.96807	0.97019	0.97218	0.97404	0.97579	0.97742	0.97895
9	0.98039	0.98173	0.98298	0.98415	0.98525	0.98627	0.98723	0.98812	0.98895	0.98973
10	0.99045	0.99113	0.99175	0.99234	0.99289	0.99340	0.99387	0.99431	0.99472	0.99510
11	0.99546	0.99579	0.99610	0.99638	0.99665	0.99689	0.99712	0.99733	0.99753	0.99771
12	0.99788	0.99804	0.99819	0.99832	0.99845	0.99856	0.99867	0.99877	0.99886	0.99895
13	0.99903	0.99910	0.99917	0.99923	0.99929	0.99935	0.99940	0.99944	0.99949	0.99953
14	0.99956	0.99960	0.99963	0.99966	0.99968	0.99971	0.99973	0.99975	0.99977	0.99979
15	0.99980	0.99982	0.99983	0.99985	0.99986	0.99987	0.99988	0.99989	0.99990	0.99991
16	0.99991	0.99992	0.99993	0.99993	0.99994	0.99994	0.99995	0.99995	0.99996	0.99996
17	0.99996	0.99997	0.99997	0.99997	0.99997	0.99998	0.99998	0.99998	0.99998	0.99998
18	0.99998	0.99999	0.99999	0.99999	0.99999	0.99999	0.99999	0.99999	0.99999	0.99999
19	0.99999	0.99999	0.99999	0.99999	1.00000	1.00000	1.00000	1.00000	1.00000	1.00000

Planetary albedo See *albedo*.

Plasma A material such as the *ionosphere* which is at a sufficiently high temperature that most of the atoms are fully ionised, so that it consists mostly of electrons and positive ions. The electromagnetic properties of a plasma are dominated by the electrons, and the *dielectric constant* is given by

$$\varepsilon_r = 1 - \frac{Ne^2}{\varepsilon_0 m\omega^2},$$

where N is the electron density (electrons per unit volume), e is the charge and m the mass of an electron, ε_0 is the permittivity of free space and ω is the angular frequency of the radiation. Electromagnetic radiation can propagate without loss through a plasma provided that ω is higher than the plasma frequency ω_p, given by

$$\omega_p = \sqrt{\frac{Ne^2}{\varepsilon_0 m}}.$$

If $\omega > \omega_p$, the *phase velocity* of electromagnetic radiation in a plasma is

$$c\left(1 - \frac{Ne^2}{\varepsilon_0 m\omega^2}\right)^{-1/2}$$

and is greater than the speed of light, c. The *group velocity* is

$$c\left(1 - \frac{Ne^2}{\varepsilon_0 m\omega^2}\right)^{1/2}$$

and is less than c.

Plasma frequency See *plasma*

PMR (Pressure-Modulated Radiometer) U.S. nadir-viewing infrared radiometer for atmospheric profiling, carried on *Nimbus-6*.

POAM-2 (Polar Ozone and Aerosol Measurement) U.S. Sun-viewing ultraviolet/ optical/near infrared limb sounder, for vertical profiling of atmospheric temperature, aerosols, H_2O, O_2, O_3 and NO_2 concentrations, carried on *SPOT-3*. Wavebands: Channels at 353, 442, 448, 600, 761, 780, 920, 936 and 1059 nm. Spectral resolution: 2 to 15 nm (filters). Observation range: Lower limit 10 km (aerosols, O_3, temperature), 15 km (H_2O), 20 km (NO_2); upper limit 40 km (60 km for O_3 and temperature).

POEM (Polar Orbiting Earth-observation Missions) *ESA*'s Polar Platform Earth Observation Mission, for operational environmental and meteorological observations, through the *Envisat* and *Metop* satellites.

POES See *NOAA-9 onwards*.

Poincaré sphere A method of representing the state of *polarisation* of completely polarised radiation. If the components of the *Stokes vector* are S_0, S_1, S_2 and S_3, the state is represented by a point on the surface of a sphere of radius S_0 centred on the origin, having Cartesian coordinates $x = S_1$, $y = S_2$, $z = S_3$. Points on the 'equator' (the xy-plane) represent linear polarisation states, the 'north pole' $(x = y = 0,\ z = +1)$ represents right-hand circularly polarised radiation, and the 'south pole' $(x = y = 0,\ z = -1)$ represents left-hand circularly polarised radiation. Other points represent elliptical polarisations.

Polarisation A restriction on the orientation of the electric field vector (or equivalently the magnetic field vector) in an electromagnetic wave. If the radiation has a single *wavenumber* k and *angular frequency* ω and is propagating in the z-direction, the electric field can be represented by its components in the x- and y-directions:

$$E_x = A_1 \cos(\phi + \delta_1),$$

$$E_y = A_2 \cos(\phi + \delta_2),$$

where A_1, A_2, δ_1, δ_2 are constants and

$$\phi = kz - \omega t.$$

In general, the path traced by the **E** vector in the xy-plane is an ellipse with semimajor axis a and semiminor axis b, oriented at an angle ψ to the x-axis, as shown below.

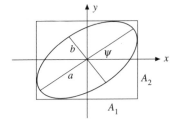

The ellipticity can be represented by the angle χ, where

$$\tan \chi = \pm \frac{b}{a}$$

and the sign is chosen according to whether the ellipse is described in a clockwise or anticlockwise direction. The parameters of the ellipse are then given by the equations

$$a^2 + b^2 = A_1^2 + A_2^2,$$

$$\tan 2\psi = \tan 2\alpha \cos \delta,$$

$$\sin 2\chi = \sin 2\alpha \sin \delta,$$

where

$$\tan \alpha = \frac{A_2}{A_1}.$$

and

$$\delta = \delta_2 - \delta_1.$$

If $\sin \delta > 0$, the radiation is said to be right-hand polarised, and if $\sin \delta < 0$ it is left-hand polarised.

A special case of elliptical polarisation occurs when $\delta = 0, \pm\pi, \pm 2\pi$ etc., in which case the radiation is **linearly polarised** (i.e. the path traced by the **E** vector in the xy-plane is a straight line). The other special case occurs when $A_1 = A_2$ and $\delta = 2\pi m \pm \pi/2$ ($m = 0, \pm 1, \pm 2$ etc.), in which case the radiation is **circularly polarised** (the path traced by the **E** vector is a circle).

If the radiation is not monochromatic, the polarisation state can change on a very short timescale. This is known as **random polarisation** (or **unpolarised** radiation), and can be analysed as two arbitrary incoherent orthogonal linear polarisations of equal amplitude. See *Stokes vector, Poincaré sphere, horizontal polarisation, parallel polarisation, perpendicular polarisation, vertical polarisation.*

Polarisation and Directionality of Earth Reflectance See *POLDER.*

Polar orbit A *low-Earth orbit* with an *inclination* close to 90°, so that the satellite passes close to the Earth's North and South Poles. A polar orbit can provide coverage of most of the Earth's surface.

Polar Ozone and Aerosol Measurement See *POAM-2.*

Polar platform See *ADEOS II, Envisat.*

Polar stereographic projection See *map projection.*

POLDER (Polarisation and Directionality of Earth's Reflectance) French optical/ near infrared *CCD* (pushbroom) imaging radiometer, with sensitivity to the state of polarisation. Carried on *ADEOS*. Wavebands: 433–453*, 480–500, 555–575, 660–680*, 745–785, 758–768, 845–885*, 900–920 nm. The bands marked with * detect three polarisation states. Spatial resolution: 6×7 km. Swath width: 2200 km.

The POLDER-II instrument is planned for inclusion on *ADEOS II*.

URL:

    http://loasys.univ-lille1.fr/recherche/pol_obs_terre/pol_obs_terre_gb.html

Population estimation In 1990 the global population was approximately 5.3×10^9, with an annual rate of growth of about 1.7%. In general, the smaller the rate of growth of a country's population, the less the difficulty of providing for the needs of the people. Increasing urbanisation and population growth impose formidable demands on the global system, because not only does the cumulative impact on humans grow, but the vulnerability of humans to various changes, climatic or otherwise, increases as well. In many countries, coastal regions are expected to experience particularly rapid growth, and this means that, if predictions of climate warming prove correct, the probable associated rise in sea level will imperil a proportionately larger number of inhabitants. Moreover, in the face of existent population pressures, few societies now have the option to avoid the damage they have inflicted on a locality simply by migrating to a virgin locality elsewhere.

The requirement by many nations for timely, accurate demographic data is recognised to be important for allocating governmental representation and also for planning purposes. Remote sensing has the potential to be valuable in estimating population. However, existing techniques need to be further developed to obtain more reliable estimates. Population estimates are inferred rather than obtained directly, by using remote sensing techniques that rely, primarily, on various land-use characteristics. Three basic approaches have been developed, all using high-resolution visible-wavelength data from aircraft or satellites.

The dwelling unit approach identifies individual dwelling units, and multiplies the number by the average family size. Deriving a representative household size is key to this approach, which is most effective in rural areas and under-developed agrarian societies.

The land use/density approach identifies residential land use, and multiplies the area represented by each type of land use by the appropriate population density. This method is more practical in developed countries where detailed GIS are available for the accurate assessment of land use.

The built-up area approach is based on the relationship between the built-up area of a settlement and its population, and is limited to more urbanised areas where the population is greater than about 500 000. A first-order determination of large-scale population patterns can be made from night-time satellite optical imagery. Many of the world's most heavily populated regions are illuminated by the bright lights of cities and towns. Indeed, the distribution of lights visible in night-time satellite imagery, especially for developed countries, often approximates the actual distribution of the population.

PoSAT-1 (Portuguese satellite) Small experimental satellite operated by Portugal and the U.K., launched in September 1993. Orbit: Circular *Sun-synchronous LEO* at 805 km altitude, inclination 98.6°, period 101.2 minutes. Principal sensors: *EIS*. The satellite also carries a *GPS* receiver for orbit determination, and a solar–terrestrial physics package.

PoSAT-1's mission was completed in January 1996.

Power pattern Characterisation of the directional properties of an *antenna*, showing how its sensitivity varies with direction. For most antennas used in remote sensing, the power pattern consists of a main beam (main lobe or major lobe), and several side lobes or back lobes. The peak value of the power pattern is unity (0 dB). The half power *beamwidth* in a given plane is the angular width between the directions at which the power pattern reaches a value of $1/2$ (-3 dB). Sometimes, other quantities may also be specified to characterise the beam, such as the null beamwidth and the −10 dB beamwidth, as shown in the figure. The power pattern is also called the radiation pattern. See also *directivity*.

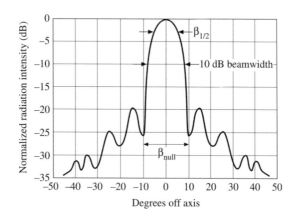

PR (Precipitation Radar) Japanese instrument for measurement of rainfall distribution, carried on *TRMM*. PR is a dual-frequency rain radar. Frequencies: K_u band (13.796, 13.802 GHz). Polarisation: HH. Spatial resolution: 4.3 km (horizontal), 250 m (range). Swath width: 220 km. Rain rate sensitivity: 0.5 mm/hr.

PRARE (Precise Range and Range-rate Equipment) German ranging system, for precise determination of satellite position, carried on *ERS-1, ERS-2* and *Meteor-3-7*. The PRARE instruments transmit signals at the S and X bands to active transponders at ground stations. The time delays of the returned signals are used to deduce the satellite's position to an accuracy of a few centimetres. An extended version, PRAREE, will use three frequencies instead of two.

Precession, orbital Rotation of the plane of a satellite's *orbit* about the Earth's polar axis, arising from the Earth's asphericity. The rate of precession Ω is given by

$$\Omega = -\frac{3J_2(GM)^{1/2}a_e^2 a^{-7/2}\cos i}{2(1-e^2)^2},$$

where J_2 is the Earth's *dynamical form factor*, G is the universal gravitational constant, M is the Earth's mass, a_e is the Earth's equatorial radius, a is the satellite's semimajor axis, i is its *inclination* and e is its *eccentricity*. The sign of Ω is positive if the precession is *prograde*, i.e. to the east. Using the best current values for J_2, G, M and a_e, the expression can be written numerically as

$$\Omega = -0.001\,319\,\frac{a^{-7/2}\cos i}{(1-e^2)^2},$$

where Ω is measured in radians per second and a is measured in thousands of kilometres.

Precipitation Precipitation, defined as liquid or solid water falling on the Earth's surface, is vital to human society, providing nearly all of the fresh water used for industry, agriculture and consumption, and strongly influences the general circulation of the atmosphere. Forecasting and monitoring of precipitation is important in numerical weather forecasting.

Although local precipitation can be measured very easily on land (using a rain gauge), synoptic data are difficult to obtain from sparsely populated land areas and over oceans. While satellite-based radiometers cannot currently measure precipitation, extensive research has shown that both rates and spatial and temporal averages can be inferred from a variety of observations. Since two thirds of the global rainfall occurs in the tropics, forecasting and monitoring of rainfall in these regions is especially important for investigations of global atmospheric circulation.

Optical ($0.3-1.0\,\mu m$) and thermal infrared ($10-12\,\mu m$) radiometers are quite successful at detecting *clouds*, and the thickness of those clouds can be inferred from their brightness and temperature. Reasonably close relationships exist between cloud presence and precipitation occurrence, and between cloud thickness and precipitation rate, and a large number of methods using these relationships to estimate totals over large areas have been used. Many of these methods have proven quite successful in estimating convective rainfall for timescales of a few days and longer and for areas of $10^4\,km^2$ and greater; the utility of optical/infrared methods for high latitude precipitation and for finer scales is unclear.

While optical and infrared radiometric measurements cannot directly detect precipitation particles, *passive microwave* observations made at frequencies from $10-200\,GHz$ can (see, for example, *SSM/I* and *MIMR*). At relatively low frequencies (below about $50\,GHz$), raindrops emit roughly as *black bodies* and can be detected over flat water surfaces which have relatively low *emissivity*. Estimates based on this mechanism can only be used over water and, because of the non-linearity of the effect, exhibit errors when the field of view of the instrument includes varying rain rates. The converse of this is that the presence of precipitation can cause difficulties in the analysis of passive microwave data, especially in discriminating *sea ice* from open water.

At higher frequencies, the dominant effect is a depression in observed temperature due to scattering by large ice particles. This depression is approximately linearly related to the density of the ice, and, in deep convective systems, to the rain rate at the surface. Estimates based on this effect can be made wherever the surface is not ice-covered, but are only useful for convective systems. Where microwave-based estimates can be made, they are quite accurate (although validation is extremely difficult), but the sampling available is limited.

Improved estimates of precipitation can be expected to come from both improved algorithms and better observations. The frequent sampling and high resolution of the operational optical and infrared radiometers, and the superior accuracy of the microwave-based methods, have recently inspired attempts to develop algorithms that use these together. Results so far are not unequivocally superior, but development is continuing. The most exciting new instrument is the *precipitation radar* on the *TRMM* satellite. However, estimates of precipitation based observations from satellite sensors will always rely on independent surface-based observations for calibration and validation. See also *rain, snow*.

Precipitation Radar See *PR*.

Precise Range and Range-rate Equipment See *PRARE*.

Preprocessing Term used in *image processing* to describe the preliminary corrections (such as *radiometric correction* and *geometric correction*) performed before information is extracted from the image.

Pressure-modulated radiometer See *PMR*.

Principal components The principal components of a multiband image are linear combinations of the bands, chosen in such a way that they are uncorrelated with each other. By convention, the order in which the principal components are specified is such that the variance decreases from the first component to the last one.

The principal components of an N-band image can be represented by

$$I'_i = \sum_{j=1}^{N} M_{ij} I_j,$$

where I_j is the digital number of a pixel in band j, and I'_i is the digital number of that pixel in principal component i. If the *covariance matrix* of the original image is C_{ij}, we can find N eigenvectors of this matrix, such that the nth eigenvector $x^{(n)}$ satisfies the equation

$$\sum_{i=1}^{N} C_{ij} x_i^{(n)} = \alpha^{(n)} x_j^{(n)},$$

where $\alpha^{(n)}$ is the corresponding eigenvalue. The eigenvectors are normalised, and arranged in order so that $\alpha^{(1)} > \alpha^{(2)} > \alpha^{(3)} \ldots$. The matrix M_{ij} defining the principal components is then given by

$$M_{ij} = x_j^{(i)}.$$

A **principal components transformation** (also known as a Karhunen–Loève transform or Hotelling transform) involves replacing the N bands of a multiband image by the corresponding N principal components. This is sometimes done when N exceeds three, so that the original data cannot all be displayed as a colour composite. The first three principal components will often contain more than 95% of the image variance, so that displaying just these three components in a false-colour image will allow most of the information present in the original image to be displayed. The principal components transformation is also used to reduce the number of bands or other features used in image *classification*.

See also *canonical components*.

Principal point The point at which a camera's optical axis intersects the ground surface. For a vertical aerial photograph, this coincides with the nadir point. See *photogrammetry*.

Priroda Remote sensing module of the Russian space station *Mir*. Became operational in 1996. 'Priroda' is Russian for 'Nature'.

Priroda-5 Russian imaging system carried on the Kristall module of the *Mir-1* space station. The system consists of two KFA-1000 cameras viewing at $\pm 8°$ from nadir. Waveband: 0.4–0.8 µm. Spatial resolution: 5 m. Image size: 100×200 km. The KFA-1000 camera has a focal length of 1000 mm and a 300 mm film size. One film can record approximately 1800 photographs.

Producer's accuracy See *error matrix*.

Prograde A prograde satellite orbit is one in which the longitudinal component of the satellite's velocity is from west to east, i.e. in the same sense as the Earth's rotation about its axis. A prograde orbit has an *inclination* of less than 90°. See *retrograde*.

Pseudocolour image See *density slicing*.

Pulse-limited Term used to describe the operation of a *radar altimeter* when the effective spatial resolution is determined by the duration of the (compressed) pulse, rather than by the *power pattern* of the antenna. The diameter of the beam-limited footprint is given by $H\Delta\theta$, where H is the range from the altimeter to the Earth's surface and $\Delta\theta$ is the power-pattern *beamwidth*. The diameter of the pulse-limited footprint is given by $2(cHt_p)^{1/2}$, where c is the speed of light and t_p is the (compressed) pulse length. The effective footprint is the smaller of the two. To date, all spaceborne radar altimeters have been pulse-limited.

Pushbroom scanner A type of *scanning system* for optical and near infrared radiation. 'Scanning' in the direction perpendicular to the motion of the platform is achieved by using a linear array of solid-state detectors (see *CCD*) onto which the radiation is focussed. The detector array contains in the order of 1000 detectors. A whole strip of pixels, extending across the entire *swath width*, is thus imaged simultaneously. Scanning in the perpendicular direction is achieved using the motion of the platform.

Compared with the *whiskbroom scanner*, the pushbroom scanner has the major advantage of having no moving parts, giving it greater robustness, lower weight and lower power consumption. It also achieves a higher signal-to-noise ratio (since it is able to sample the radiation from a given resolution element for a longer period of time) and a higher geometric accuracy. The disadvantage of the pushbroom scanner is the much larger number of detectors to be calibrated. Compare also *step-stare imager*.

PVI (Perpendicular vegetation index) See *vegetation index*.

Q band Subdivision of the microwave region of the *electromagnetic spectrum*, covering the frequency range 36 to 46 GHz (wavelengths 6.5 to 8.3 mm).

Quality of life Quality of life is experienced subjectively and so is not an entity that can easily be quantified or even observed. However, certain measures which relate to quality of life, and which certainly affect humans, and especially the potential for human suffering, can be monitored from space. It is very difficult to monitor those things that make one's life 'good' since they may vary from culture to culture, and in some cases are abstract feelings rather than physical features. We may be able to surmise that the quality of life in a remote village is high if the crops appear healthy and the water supply is not laden with sediment or chemicals. But it is far easier to postulate about the human condition by observing what people must endure in regard to both natural and man-made or induced disasters. For example, what appears to be an unmitigated disaster may actually be a blessing in disguise. Floods that endanger riparian settlements also provide nutrients to replenish spent soils. In the long run, an event such as this may make a place better to live in, but in the short term, loss of material possessions, shelter, and the calamity that follows, may make the lives of the survivors truly miserable.

Perhaps it is the human-induced changes that have the largest impact on quality of life. Causing our own undoing or suffering, such as in the case of the Chernobyl nuclear power plant accident, is particularly egregious, because, unlike with natural disasters, faulty decision-making is responsible for or leads to a reduced quality of life. In view of the widespread changes that human populations are causing to the environment, and increased awareness of these changes, it is timely that spaceborne sensors are now available to monitor them. The information that these sensors reveal can contribute to policy decisions that will help reverse detrimental changes. Of course, not all human-induced changes are negative, and the hope must be that in the coming decades spaceborne sensors, in addition to monitoring environmental damage, will also be recording improvements resulting from increased awareness and concern.

See also *disaster monitoring*.

QuickBird U.S. satellite planned for launch in September 1998, operated by *EarthWatch Inc.*, with a design life of 5 years. <u>Objectives</u>: High-resolution

Earth observation for commercial use. Orbit: Circular *LEO* at 600 km altitude. Inclination 52°.

QuickBird will carry high-resolution optical/near infrared pushbroom (CCD) imagers, providing panchromatic and multispectral imagery. Wavebands: 0.45–0.90 µm (panchromatic); 0.45–0.52, 0.52–0.60, 0.63–0.69, 0.76–0.90 µm (multispectral). Spatial resolution: 0.82 m (panchromatic); 3.3 m (multispectral). Image size: 22 × 22 km.

The QuickBird sensor's field of view can be steered by ±30° fore and aft and side to side, to provide stereo imaging and greater flexibility in coverage.

URL: http://www.digitalglobe.com/quickbird/qbov.html

Quick look Quick-look images are low-resolution hard-copy or digital versions of remote sensing images, used for confirming the suitability of an image (for example, in terms of coverage, cloud cover etc.) before purchasing it. On-line internet catalogues are now available for the most commonly used imagery, for example at:

Earthnet	http://earthnet.esrin.esa.it/
EROS Data Center	http://edcwww.cr.usgs.gov/webglis
EWSE	http://ewse.ceo.org/anonymous/indexpage/quicklooks.html
Restec (Japan)	http://hgssac01.eoc.nasda.go.jp/~goin/index.html
SPOT Image	http://www.spotimage.fr/anglaise/offer/catalo/oc_consu.htm

R225 Russian single-frequency nadir-viewing passive microwave radiometer, planned for inclusion on *Okean-O* satellites. <u>Frequency</u>: 13.3 GHz. <u>Spatial resolution</u>: 130 km.

R400 Russian single-frequency passive microwave radiometer, carried on Priroda module of *Mir-1* space station. <u>Incidence angle</u>: 40°. <u>Frequency</u>: 7.5 GHz.

R600 Russian single-frequency nadir-viewing passive microwave radiometer, planned for inclusion on *Okean-O* satellites. <u>Frequency</u>: 5 GHz. <u>Spatial resolution</u>: 130 km.

RA See *radar altimeter*.

Radar (RAdio Detection And Ranging) General term to describe *active* remote sensing in the *microwave* region of the electromagnetic spectrum. *Synthetic aperture radars* and *side-looking radars* provide the two-dimensional coverage necessary for the generation of images; *radar altimeters* and *scatterometers* are non-imaging systems.

Radar systems transmit microwave radiation either continuously or as short pulses, pulsed operation being more common in remote sensing applications. The received signal is analysed for intensity (or amplitude and phase in the case of coherent systems such as synthetic aperture radar), time delay, and possibly also *polarisation* state and/or *Doppler* shift.

See *radar equation*.

Radar altimeter

1. A non-imaging instrument, primarily used for measuring the distance from the instrument to the Earth's surface. A radar altimeter emits a very short pulse of microwave radiation, and measures the time taken before the signal reflected from the Earth's surface is received. If the propagation speed of the signal through the Earth's atmosphere is known (see *tropospheric delay*, *ionosphere*), the range can be calculated, and if the position of the satellite carrying the altimeter is known (for example by laser ranging or from an on-board *GPS* receiver), the absolute location of the reflecting point can be deduced.

The extremely short pulses used by radar altimeters are in fact usually generated synthetically, using a 'chirped' pulse in which the frequency ν varies linearly with time from $\nu_0 - \Delta\nu/2$ to $\nu_0 + \Delta\nu/2$ over a time T_{p}, where ν_0 is the carrier frequency. With appropriate signal processing, such a chirped signal is equivalent to a pulse of effective duration $t_p = 1/\Delta\nu$. A typical system might have the following parameters: $\nu_0 = 13.6\,\text{GHz}$, $T_p = 20\,\mu\text{s}$, $\Delta\nu = 300\,\text{MHz}$, giving an effective value of $t_p = 3.3\,\text{ns}$. If the effective duration of the emitted pulse is t_p, the range to a smooth surface can be determined to a precision of approximately $ct_p/2$ from a single measurement, where c is the speed of light. Since this precision can be of the order of 1 metre or less, and is often improved by averaging successive pulses, corrections for phase delays in the ionosphere and atmosphere must be made if the errors arising from these sources are not to dominate the accuracy of the range measurement. The spatial resolution of a radar altimeter, typically in the range 1 to 10 km, depends on whether the instrument is *pulse-limited* or *beam-limited*.

Radar altimeter measurements have been extensively used for measuring the topography of ocean, ice sheet and large lake surfaces. In the case of oceans, the mean *sea surface topography* (with the effects of tidal variations averaged out) has been used to define the *geoid* and the sea-floor topography. Information on surface roughness within the footprint of the radar altimeter is also available, if the instrument records the variation with time of the returned signal. This is because a rough surface will cause the return signal to be spread out over a time of the order of $\Delta h/c$, where Δh is the r.m.s. variation of the surface height. This technique is used to characterise sea-state by measuring the *significant wave height*, and hence to infer *wind speed* over oceans. Radar altimeter measurements have also been used to determine the extent and freeboard of *sea ice*.

See also *electromagnetic bias*, *slope correction*.

Table of principal satellite radar altimeters.

Satellite	Launch year	Frequency (GHz)	t_p (ns)	Height (km)	Footprint Beam (km)	Footprint Pulse (km)
GEOS-3	1975	13.9	12.5	843	38	3.5
Seasat	1978	13.5	3.1	800	22	1.7
Geosat	1985	13.5	3.1	800	29	1.7
ERS	1991	13.8	3.0	800	18	1.7
Topex-Poseidon	1992	13.6/5.3	3.1	1300	26/65	2.2
Mir-1 (Greben)	1993	13.8	12.5	400	13	2.3
Envisat	1999	13.6/3.2	3.1/6.3	800	26/90	1.7/2.5
Jason-1	1999	5.3/13.8	3.1	1334	450	2.2

2. Generic name used for the radar altimeters described below.

Nadir-viewing instruments on **ERS-1** and **ERS-2**. Frequency: K_u band (13.8 GHz). Pulse length (uncompressed): 20 µs; (compressed): 3.0 ns ('ocean mode'), 12.1 ns ('ice mode'). Pulse repetition frequency: 1020 Hz. Beam-limited footprint: 18 km. Pulse-limited footprint: 1.7 km (ocean mode), 3.4 km (ice mode).

The ERS altimeters can be switched from 'ocean mode' to the lower-resolution 'ice mode' which is better able to maintain tracking over comparatively steeply sloping margins of ice sheets.

An improved, dual-frequency version, RA-2, is planned for inclusion on *Envisat*. Frequencies: K_u band (13.6 GHz), S band (3.2 GHz). Pulse length (uncompressed): 20 μs; (compressed): 3.1/12.5/50 ns (K_u), 6.3 ns (S). Pulse repetition frequency: 1800 Hz (K_u), 450 Hz (S). Beam-limited footprint: 26 km (K_u), 90 km (S). Pulse-limited footprint: 1.7/3.5/6.9 km (K_u), 2.4 km (S).

The dual-frequency operation allows ionospheric correction. Programmable range resolution allows different surfaces to be tracked. 100 K_u-band and 25 S-band pulses are averaged on board.

Nadir-viewing instrument on *GEOS-3*. Frequency: K_u band (13.9 GHz). Pulse length (uncompressed): 1.0 μs; (compressed): 12.5 ns. Height precision: 50 cm ('global mode'), 20 cm ('intensive mode'). Pulse repetition frequency: 1 Hz. Beam-limited footprint: 38 km. Pulse-limited footprint: 3.5 km.

Absolute altitudes from the GEOS-3 altimeter are subject to unmeasured propagation errors.

Nadir-viewing instrument on *Geosat*. Frequency: K_u band (13.5 GHz). Pulse length (uncompressed): 102 μs; (compressed): 3.1 ns. Height precision: 5 cm. Pulse repetition frequency: 10 Hz. Beam-limited footprint: 29 km. Pulse-limited footprint: 6.8 km. Processed data have an along-track spacing of 6.7 km.

Nadir-viewing instrument on *GFO*. Frequency: K_u band (13.5 GHz). Height precision: 3.5 cm. Pulse-limited footprint: 2 km.

Nadir-viewing instrument on *Seasat*. Frequency: K_u band (13.5 GHz). Height precision: 10 cm. Pulse length (uncompressed): 3.2 μs; (compressed): 3.1 ns. Beam-limited footprint: 22 km. Pulse-limited footprint: 1.7 km.

Radar equation Equation relating the power P_r received by a *radar* to the transmitted power P_t, the viewing geometry, and the properties of the scattering surface. The **bistatic** radar equation is used when different antennas or geometries are used for transmitting and receiving. It can be written as

$$P_r = \frac{P_t G_t A_r}{16\pi^2 R_t^2 R_r^2} \sigma^0 \, dA,$$

where G_t is the *gain* of the transmitting *antenna*, A_r is the *effective area* of the receiving antenna, R_t and R_r are the distances from the transmitting and receiving antennas, respectively, to the scattering surface, dA is the area from which radiation is being scattered, and σ^0 is its *backscatter coefficient* for the particular frequency, polarisation and scattering geometry. The **monostatic** radar equation is a special case of this formula, corresponding to the common situation in which the same antenna is used, with the same geometry, for transmitting and receiving radiation. In this case,

$$R_t = R_r = R$$

and

$$G_t = \frac{4\pi A_r}{\lambda^2} = G,$$

where λ is the wavelength, so that

$$P_r = \frac{P_t \lambda^2 G^2}{(4\pi)^3 R^4} \sigma^0 \, dA.$$

Radargrammetry The process of generating a planimetric map from imaging *radar*. The technique is similar in principle to *photogrammetry*, although the oblique incidence angle and geometric distortions inherent in radar images impose significant difficulties.

Radarsat Canadian satellite, operated by *CSA*, launched in November 1995 with a nominal lifetime of 5 years. <u>Objectives</u>: Arctic sea ice, agriculture, forestry and oceanographic observations. Data continuity with *ERS-1* and *JERS-1*. <u>Orbit</u>: Circular *Sun-synchronous LEO* at 790 km altitude. Period 100.8 minutes; inclination 98.6°; equator crossing time 18:00 (ascending node). *Exactly repeating orbit* (343 orbits in 24 days). <u>Principal instruments</u>: *SAR*.

URL: http://www.rsi.ca/

Radar transponder A device that collects microwave radiation and retransmits it, possibly after amplification, back in the direction from which it was received. Radar transponders can be used to provide *ground control points* for radar imagery or for *radar altimeter* measurements. They are also included on some satellites so that the satellite's position can be determined by radar ranging from ground stations.

The simplest type of radar transponder is the **passive transponder**, which merely reflects the signal without amplification. A common design is the **corner-cube reflector**, which consists of three mutually perpendicular reflecting planes (usually metal sheets or wire grids) in the form of 45°–45°–90° triangles, as shown below.

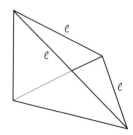

The maximum backscatter cross-section σ (see *backscatter coefficient*) of the corner-cube reflector is approximately ℓ^4/λ^2, where ℓ is the length of the hypotenuses of the triangles and λ is the wavelength of the radiation. This maximum value is obtained in the direction along the axis of three-fold rotational symmetry of the reflector. The half-power *beamwidth* of the device is approximately $30°$.

If a greater backscatter cross-section is required, an **active transponder** can be used. This consists essentially of a transmitting and receiving antenna and an amplifier.

Radiance The contribution to the electromagnetic power falling on a plane surface of area dA from a solid angle $d\Omega$ in a direction making an angle θ to the normal to the area dA is given by

$$L \cos \theta \, d\Omega \, dA,$$

where L is the radiance from that direction. The SI unit of radiance is $W \, m^{-2} \, sr^{-1}$. For isotropic radiation the radiance is the same in all directions.

Radiant exitance The total electromagnetic power emitted in all directions by a plane surface area dA is defined by

$$M \, dA,$$

where M is the radiant exitance. The SI unit of radiant exitance is $W \, m^{-2}$. If the outgoing *radiance* from the surface is $L(\theta, \phi)$ in the direction which makes an angle θ with the surface normal and has azimuth angle ϕ, the radiant exitance is given by

$$M = \int_{\theta=0}^{\pi/2} \int_{\phi=0}^{2\pi} L(\theta, \phi) \sin \theta \, d\theta \, d\phi.$$

Radiant flux The radiant flux onto a body is the total electromagnetic power incident upon it. It is given by

$$\Phi = \int E \, dA,$$

where E is the *irradiance* on surface area dA. Similarly, the radiant flux leaving a body is given by

$$\Phi = \int M \, dA,$$

where M is the *radiant exitance* from surface area dA. The SI unit of radiant flux is W.

Radiant intensity The radiant intensity of electromagnetic radiation is the *radiant flux* per unit solid angle. The SI unit of radiant intensity is $W \, sr^{-1}$.

Radiation pattern See *power pattern*.

Radiative transfer equation Equation describing the propagation of electro-magnetic radiation in the presence of absorption, *scattering* and *black-body* emission of radiation. If the radiation is in thermal equilibrium with the medium through which it propagates, the equation can be written as

$$\frac{dL_\nu(\theta, \phi)}{dz} = -\gamma L_\nu(\theta, \phi) + \gamma_a B_\nu + \gamma_s J_\nu,$$

where $L_\nu(\theta, \phi)$ is the spectral *radiance* of the radiation propagating in the direction (θ, ϕ) and B_ν is the spectral radiance of black-body radiation corresponding to the temperature T of the medium, given by

$$B_\nu = \frac{2h\nu^3}{c^2} \frac{1}{e^{h\nu/kT} - 1}.$$

γ_a is the *absorption coefficient*, γ_s is the *scattering coefficient*, $\gamma = \gamma_a + \gamma_s$ is the *attenuation coefficient*, and z measures distance in the direction of propagation. J_ν describes the radiation scattered into the direction (θ, ϕ) from directions (θ', ϕ') and is given by

$$J_\nu = \frac{1}{4\pi} \int_{4\pi} L_\nu(\theta', \phi') p(\cos\Theta) \, d\Omega',$$

where $p(\cos\Theta)$ is the scattering **phase function**, which describes the angular distribution of the scattered radiation, and

$$\cos\Theta = \cos\theta \cos\theta' + \sin\theta \sin\theta' \cos(\phi - \phi').$$

$d\Omega' = \sin\theta' d\theta' \, d\phi'$ is an element of solid angle, and the integration is performed over all directions (i.e. 4π steradians).

Various special cases of the radiative transfer equation are useful in remote sensing applications:

1. Absorption only
Setting $\gamma_s = 0$ and $B_\nu = 0$ gives

$$\frac{dL_\nu}{dz} = -\gamma_a L_\nu$$

(the direction (θ, ϕ) is implicit), which has the solution

$$L_\nu = L_0 e^{-\tau},$$

where L_0 is a constant and

$$\tau = \int \gamma_a \, dz$$

is the *optical thickness* of the medium (see *Lambert–Bouguer law*).

2. Absorption and emission only
Setting $\gamma_s = 0$ gives

$$\frac{dL_\nu}{dz} = \gamma_a (B_\nu - L_\nu)$$

(again, the direction (θ, ϕ) is implicit).

2.1. Constant B_ν

For constant B_ν, the solution is

$$L_\nu = L_0\, e^{-\tau} + B_\nu(1 - e^{-\tau}),$$

where L_0 is a constant and τ is again the optical thickness. If the frequency ν is low enough for the *Rayleigh–Jeans approximation* to be valid, the solution can be rewritten as

$$T_{\mathrm{b,out}} = T_{\mathrm{b,in}}\, e^{-\tau} + T(1 - e^{-\tau}),$$

where $T_{\mathrm{b,out}}$, is the *brightness temperature* of the radiation emerging from an absorbing medium of optical thickness τ and physical temperature T, and $T_{\mathrm{b,in}}$, is the brightness temperature of the radiation entering the medium.

2.2. Varying B_ν

Consider an absorbing and emitting medium extending from $z = 0$ to $z = z_0$, with radiation of spectral radiance L_{in} incident in the $+z$ direction at $z = 0$. The solution of the radiative transfer equation is most simply expressed if position z within the medium is expressed in terms of the *optical depth* $\tau(z)$, defined as the optical thickness between z and z_0:

$$\tau(z) = \int_z^{z_0} \gamma_a(z')\, dz',$$

such that $\tau(z_0) = 0$. The value of $\tau(0)$, the optical thickness of the entire medium, is written as τ_0 (see figure).

The spectral radiance of the emerging radiation is then

$$L_{\mathrm{out}} = L_{\mathrm{in}}\, e^{-\tau_0} + \int_0^{\tau_0} B_\nu(\tau')\, e^{-\tau'}\, d\tau'.$$

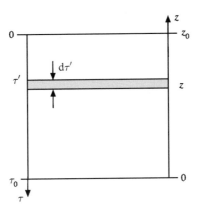

The optical depth τ' is the optical thickness between z and z_0. The contribution to the spectral radiance propagating in the $+z$ direction from the layer of thickness $d\tau'$ is $B_\nu(\tau') \exp(-\tau')\, d\tau'$.

This formula is widely applied in modelling atmospheric temperature sounding observations. At low enough frequencies in the microwave region, where the Rayleigh–Jeans approximation is valid, it can be rewritten as

$$T_{b,out} = T_{b,in}\, e^{-\tau_0} + \int_0^{\tau_0} T(\tau')\, e^{-\tau'}\, d\tau'.$$

3. Absorption and scattering only

If $B_\nu = 0$, the radiative transfer equation becomes

$$\frac{dL_\nu(\theta, \phi)}{dz} = -\gamma L_\nu(\theta, \phi) + \frac{\gamma_s}{4\pi} \int_{4\pi} L_\nu(\theta', \phi') p(\cos\Theta)\, d\Omega'.$$

This equation is widely used in modelling *volume scattering* phenomena.

3.1. One-dimensional case

The one-dimensional simplification of this equation can be written as

$$\frac{dI_+}{dz} = -\gamma I_+ + \gamma_s I_-$$

and similarly

$$\frac{dI_-}{dz} = -\gamma_s I_+ + \gamma I_-,$$

where I_+ is the intensity of radiation propagating in the $+z$ direction and I_- is the intensity in the $-z$ direction. If γ and γ_s are constant, these equations have the solutions

$$I_+ = I_{+,0}\, e^{-\mu z},$$

$$I_- = I_{-,0}\, e^{-\mu z},$$

where $I_{+,0}$ and $I_{-,0}$ are constants, and

$$\mu^2 = \gamma^2 - \gamma_s^2 = \gamma_a(\gamma_a + 2\gamma_s).$$

Radiative transfer models Radiative transfer models are volume *scattering models* developed from the *radiative transfer equation*. This formulation uses the intensity rather than the scattered field, and states that the intensity propagating along a path is decreased due to extinction loss and increased due to scattering into the path. This formulation is useful for computing volume scattering from a discrete random medium and is capable of including polarisation and multiple-scatter effects. The scattering by a local scatterer is represented by a scattering phase function which is usually calculated from field considerations in the microwave region and modelled from selected functions in the optical region. Multiplying the phase function by the number density of scatterers yields the scattering coefficient per unit volume, which is the quantity needed in the radiative transfer formulation.

It is possible to derive a volume scattering model for a forest canopy, sea ice, snow etc. by choosing different phase functions for different types of scatterer.

The use of such a phase function implies that each scatterer scatters independently of others, and this is true in a sparsely populated medium. In a densely populated medium one must derive the phase function for the group of scatterers within unit volume, because the concept of the number density of scatterers no longer applies. The major disadvantage of this approach is that it cannot account for phase interference effects in multiple scattering calculations. It is, however, presently the only practical way to calculate multiple scattering effects.

See *dense medium models*.

Radiometer A non-imaging sensor used to measure the intensity of electromagnetic radiation in some waveband in any part of the electromagnetic spectrum (e.g. optical, infrared, microwave). A radiometer operating in visible to thermal infrared wavelengths will be used to record optical radiance properties such as radiant flux, irradiance or radiance that describe the optical properties of a target.

An optical/infrared radiometer comprises three elements: an optical system of lenses, mirrors, apertures, modulators and dispersion devices; detectors that produce an electrical signal proportional to the intensity of the incoming radiation; a signal processor that converts the electrical signal into the desired output data (single values of spectral radiance, or a spectrum). Many radiometers are now available, varying in sophistication from a dual-waveband analogue hand-held radiometer used to measure the *photosynthetically active radiation* reflected or transmitted by a crop canopy, to a digital, satellite-borne spectrometer used for atmospheric sounding. The major technical characteristics that determine the utility of a radiometer for a particular application are its spectral range, spectral resolution and wavelength accuracy, radiometric resolution and signal-to-noise ratio, calibration method and stability, field-of-view, and speed of measurement. Today, radiometers are used to understand the interaction of electromagnetic radiation with a wide range of target materials (oceans, forests, ice sheets etc.), to infer the characteristics of targets (e.g. crop biomass, soil moisture, water turbidity, air temperature), and to aid in the interpretation of imagery recorded by electro-optical sensors such as *scanners*.

Radiometric correction The process of converting the intensity of the signal measured by a remote sensing instrument into the appropriate physical units, such as units of *radiance*. Full radiometric correction requires: (1) *calibration* of the sensor; (2) *atmospheric correction*; and (3) in the case of observations made at a range of incidence angles, a model of the target reflectance or backscattering properties. Satellite-based instruments are normally calibrated before launch, but the calibration may change over time. Some instruments are provided with the means of recalibration during operation, by periodically observing on-board calibrators or (for example) the Sun. Caution should be exercised in using data from instruments that do not have this facility.

Radiometric data from most commercially available instruments are available in calibrated form, usually by rescaling the *digital numbers* so that they are linearly dependent on radiance, radar backscatter amplitude etc.

Radiometric resolution The ability of a sensor to discriminate between different values of signal intensity. The radiometric resolution is limited fundamentally by the signal-to-noise ratio, but may be further degraded by the process of digitising (quantising) the detected signal. Since the signal presented to a detector can be increased by acquiring it from a larger spatial region or a wider spectral bandwidth, radiometric resolution can be increased at the expense of *spatial resolution* or *spectral resolution*.

The radiometric resolution of a *passive microwave radiometer* is the standard deviation associated with the *antenna temperature*. For most radiometer receivers, the radiometric sensitivity ΔT may be defined as

$$\Delta T = \frac{M(T_A + T_{\text{rec}})}{\sqrt{\tau \Delta \nu}},$$

where M is the figure of merit associated with the type of receiver used, T_A is the antenna temperature at the antenna terminals, T_{rec} is the receiver noise temperature, τ is the integration time of the receiver and $\Delta \nu$ is the receiver bandwidth. For a total-power receiver, $M = 1$, while for a Dicke-switched receiver, $M = 2$. Radiometers used in remote sensing applications typically have $\Delta T \approx 1$ K.

Rain, attenuation of microwaves by Approximate values of the *attenuation coefficient* in dB/km as a function of rain rate R (mm/hr) and frequency f (GHz) are given in the table below.

R	f	3	5	10	20	30	50	100
0.25						0.05	0.12	0.35
1.25					0.15	0.3	0.8	2
5				0.09	0.4	1.0	2	5
10				0.2	1.0	2.0	4	7
25			0.1	0.8	3	5	8	12
50			0.2	1.5	5	10	17	20
100		0.05	0.5	4	12	20	28	28
200		0.09	0.9	8	30	40		

See also *precipitation*.

Rain radar See *PR*.

Random polarisation See *polarisation*.

Range direction The direction on the Earth's surface perpendicular to the motion of a *side-looking radar* or *synthetic aperture radar*, also called the across-track direction.

RAR See *side-looking radar*.

Raster format Representation of spatial data, such as a map or image, by a grid of *pixels*. Compare *vector format*. Raster format is also known as grid format.

Ratio vegetation index See *vegetation index*.

Rayleigh criterion The conventional definition of a specularly smooth surface (one that will give *specular reflection* for radiation of wavelength λ) is one which satisfies the Rayleigh criterion

$$\Delta h < \frac{\lambda}{8 \cos \theta},$$

where Δh is the root-mean-square variation of the surface height and θ is the angle between the incident radiation and the surface normal.

Rayleigh distribution See *speckle*.

Rayleigh–Jeans approximation An approximation to the *Planck distribution*, increasingly accurate at lower frequencies. The *spectral radiance* of a black body can be written as

$$L_\nu \approx \frac{2kT\nu^2}{c^2} = \frac{2kT}{\lambda^2},$$

where k is Boltzmann's constant, T is the absolute temperature, c is the speed of light, ν is the frequency and λ is the wavelength. The approximation is accurate to better than 1% for $\nu < 0.42T$, to better than 0.1% for $\nu < 0.042T$ and so on, where ν is measured in GHz and T in K.

Rayleigh scattering *Scattering* of radiation by particles small compared with the wavelength. For a spherical particle of *refractive index* n and radius a, the scattering cross-section σ_s, defined as the ratio of the scattered power to the incident radiation flux, is given by

$$\sigma_s = \frac{128\pi^5 a^6}{3\lambda^4} \left| \frac{n^2 - 1}{n^2 + 2} \right|^2,$$

where λ is the wavelength. The absorption cross-section σ_A is given by

$$\sigma_A = \frac{8\pi^2 a^3}{\lambda} \, \mathrm{Im}\left(-\frac{n^2 - 1}{n^2 + 2} \right).$$

These formulae are in general valid only when $\alpha < \lambda/100$. For larger particles, the formulae for *Mie scattering* must be used.

RBV (Return-beam vidicon) U.S. optical/near infrared *vidicon* imaging radiometer system, carried on *Landsat-1* to *-3* satellites. Wavebands: 0.48–0.58, 0.58–0.68, 0.70–0.83 μm. Spatial resolution: 79 m (Landsat-1 and -2), 40 m (Landsat-3). Swath width: 185 km.

Receiving station A facility for receiving spaceborne remote sensing data down-linked from the satellite carrying the remote sensing instrument as a modulated radio signal. The data can be downlinked directly to Earth in real time, or stored on board and then downlinked to a receiving station when the satellite is within the *station mask*. Alternatively, the data can be transmitted to a *relay satellite* and then downlinked to a receiving station. See *ground segment*.

Red edge The sharp increase in the spectral reflectance of *vegetation* occurring at a wavelength of about 0.75 μm. See *vegetation index*.

Red–Green–Blue display See *RGB display*.

Reflective infrared See *infrared*.

Reflectivity See *bidirectional reflectance distribution function*.

Refractive index The ratio of the speed of electromagnetic radiation *in vacuo* to the *phase velocity* of radiation in a medium. It is usually given the symbol n, and is related to the *dielectric constant* ε_r and relative magnetic permeability μ_r (which has a value close to 1 for most materials of interest in remote sensing) of the medium by

$$n = \sqrt{\varepsilon_r \mu_r}.$$

The refractive index can be complex, in which case it can be written as

$$n = n' - i\kappa,$$

where $i^2 = -1$. If $\mu_r = 1$ and $\varepsilon_r = \varepsilon' - i\varepsilon''$,

$$n' = \sqrt{\frac{\varepsilon' + \sqrt{\varepsilon'^2 + \varepsilon''^2}}{2}},$$

$$\kappa = \frac{\varepsilon''}{\sqrt{2(\varepsilon' + \sqrt{\varepsilon'^2 + \varepsilon''^2})}}.$$

For radiation with an angular frequency ω, this represents a wave with a phase velocity c/n' and an amplitude which decreases exponentially with distance travelled, as a result of absorption in the medium, such that it is reduced to a factor of $1/e$ of its initial value in a distance $c/\omega\kappa$. The *absorption coefficient* γ_a is given by

$$\gamma_a = \frac{2\omega\kappa}{c}.$$

In these formulae, c is the speed of light *in vacuo* and e is the base of natural logarithms.

Relief displacement In a vertical (nadir-looking) aerial photograph or satellite image, the position of the image of an object depends not only on the object's horizontal coordinates but also on its height. Higher objects are displaced further from the centre of the photograph. This phenomenon is called relief displacement, and forms the basis of height determinations from *stereophotography*.

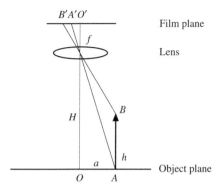

The diagram shows a vertical object AB of height h located a distance a from the nadir point O. The camera lens is a height H above the object (ground) plane. The relief displacement $A'B'$ is given by

$$\frac{hfa}{H(H-h)},$$

where f is the focal length of the camera lens.

See also *photogrammetry*.

Repeat period The interval between successive passages of a satellite's *sub-satellite point* through the same point on the Earth's surface, in the same direction. See *exactly-repeating orbit*.

Resampling The process of assigning *digital numbers* (DNs) to the pixels in an image that has been spatially transformed by *geometric correction*, using the digital numbers of the original (untransformed) image as input. In general, the transformation is described by a pair of mapping functions

$$u = f(x, y),$$

$$v = g(x, y),$$

and their inverses

$$x = F(u, v),$$

$$y = G(u, v),$$

where (x, y) are the coordinates of a point in the original image and (u, v) are the coordinates of the corresponding point in the transformed image. If the units of measurement for x, y, u and v are the dimensions of a pixel, all the coordinates should be integers. The problem is that the inverse mapping functions F and G will not necessarily produce integer values of x and y for integer values of

u and v, so that no single pixel in the original image corresponds to the pixel (u, v) in the transformed image.

Nearest-neighbour resampling simply assigns to pixel (u, v) in the transformed image the DN of the pixel in the original image that is geometrically closest to the calculated position (x, y). This can be expressed as

$$I'(u, v) = I([x], [y]),$$

where $I'(u, v)$ is the DN assigned to pixel (u, v) in the transformed image, I is the DN in the original image, $[x]$ is defined as the integer closest to x (and similarly for $[y]$), $x = F(u, v)$ and $y = G(u, v)$.

Interpolation-based resampling uses the DN values of pixels near to (x, y), and a simple model of their spatial variation, to estimate $I'(u, v)$. The simplest of these is **bilinear interpolation**, which uses four neighbouring pixels:

$$I'(u, v) = aI(x_0, y_0) + bI(x_0 + 1, y_0) + cI(x_0, y_0 + 1) + dI(x_0 + 1, y_0 + 1),$$

where x_0 is the largest integer not exceeding x and y_0 is the largest integer not exceeding y. The coefficients a, b, c and d are given by

$$a = (x_0 + 1 - x)(y_0 + 1 - y),$$
$$b = (x - x_0)(y_0 + 1 - y),$$
$$c = (x_0 + 1 - x)(y - y_0),$$
$$d = (x - x_0)(y - y_0).$$

Cubic interpolation (cubic convolution) uses 16 neighbouring pixels:

$$I'(u, v) = \sum_{i=-1}^{2} \sum_{j=-1}^{2} a_{ij} I(x_0 + i, y_0 + j).$$

The coefficients a_{ij} are given by

$$a_{ij} = h_i(x)h_j(y),$$

where

$$h_{-1}(x) = -\frac{x'}{3} + \frac{x'^2}{2} - \frac{x'^3}{6},$$

$$h_0(x) = 1 - \frac{x'}{2} - x'^2 + \frac{x'^3}{2},$$

$$h_1(x) = x' + \frac{x'^2}{2} - \frac{x'^3}{2},$$

$$h_2(x) = -\frac{x'}{6} + \frac{x'^3}{6},$$

and $x' = x - x_0$, and similarly for $h_i(y)$.

Interpolation methods give smoother images than nearest-neighbour resampling, but at the expense of potentially introducing artefacts into the data. The artefacts can be reduced by using a larger number of neighbouring pixels, but this, of course, is at the expense of a longer processing time. Nearest-neighbour resampling, which does not introduce artefacts and merely rearranges the pixels spatially, is thus generally used if the data are to be classified.

Resolution The fineness with which an instrument can distinguish between different values of some measured property. In remote sensing applications, three important types of resolution can be identified: *spatial resolution*, which refers to the spatial extent of the smallest resolvable element (e.g. a *pixel*); *spectral resolution*, which refers to the smallest wavelength or frequency change detectable in a spectral measurement; and *radiometric resolution*, which refers to the amplitude or intensity of the detected signal. Technical considerations normally cause these three types of resolution to be inter-related, so that one can be improved at the expense of the others.

Temporal resolution relates to the time interval between successive opportunities to observe a given location. For a spaceborne system, this is determined by the satellite *orbit*, the *swath width* of the instrument, and whether the instrument has a fixed or variable viewing geometry.

Resurs-F Series of Russian remote-sensing spacecraft carrying photographic cameras on short (two to four weeks) missions, from 1994 onwards. Data are retrieved from the recoverable Vostok capsule. Orbit (typical): *semimajor axis* 6640 km to 6710 km, *inclination* 82.5°, *period* 90 to 91 minutes. Typical instruments: *KFA-200, KFA-1000, MK-4*. Resurs-F3 will carry two KFA-3000 cameras giving a spatial resolution of 2 m.

Resurs-O Series of Russian (formerly Soviet) remote-sensing satellites based on the *Meteor* spacecraft; similar in concept to the U.S. *Landsat* programme. The first satellite in the series, Resurs-O1-1, was launched in October 1985. The earlier satellites (Resurs-O1) had nominal lifetimes of 2 years; Resurs-O2 satellites (1995 onwards) have nominal lifetimes of 3 years. Objectives: Environmental monitoring and measurement. Orbit: Nominally circular *Sun-synchronous LEO* at 650 km altitude (835 km from 1996 onwards). Period 98 minutes; inclination 98° (99° from 1996 onwards). Principal instruments: *MSU-E, MSU-SK, SAR* (Resurs-O2 series).

URL: http://www.eurimage.it/Products/RESURS_O1.html

Retrograde A retrograde satellite orbit is one in which the longitudinal component of the satellite's velocity is from east to west, i.e. in the opposite sense to the Earth's rotation about its axis. A retrograde orbit has an *inclination* of more than 90°. Compare *prograde*.

Retroreflector See *laser retroreflector*.

Return-beam vidicon See *RBV*.

Rezel (Resolution element) The area on the Earth's surface corresponding to a *pixel* in an image.

RGB display The usual form of colour display for a digital image, in which three channels of data (e.g. three spectral bands) are displayed in the three primary colours red, green and blue, for example, by being applied to the red, green and blue guns of a colour cathode ray tube. The colours are combined additively, so that maximum values in each channel will combine to produce white. See also *IHS display*.

RLSBO Ukrainian *SLAR* system carried on *Okean-O-1* series and planned for inclusion on *Okean-O*. Frequency: X band (9.7 GHz). Polarisation: VV. Incidence angle: 20° at near edge of swath. Spatial resolution: 2.1–2.8 km (azimuth), 1.2–0.7 km (range). Swath width: 450 km (two instruments viewing left and right of the sub-satellite track).

RM-08 Ukrainian scanning passive microwave radiometer for surface and atmospheric observations, carried on *Okean-O1* satellites. Frequency: 36.6 GHz. Spatial resolution: 15 × 20 km. Swath width: 550 km.

Roberts operator See *edge detection*.

Rocks and minerals, electromagnetic properties The data listed below summarise the electromagnetic properties of selected rock types.

Visible/near infrared region
Many rocks and minerals have characteristic spectral reflectance properties in the visible and near infrared region of the electromagnetic spectrum, often as a consequence of particular absorption lines. The table illustrates some typical spectral reflectances in per cent between 0.4 and 1.5 µm ('w' denotes weathered samples). A large number of high-resolution spectra of minerals are maintained by the United States Geological Survey and are accessible at the following URL: http://speclab.cr.usgs.gov/index.html

Material	Wavelength (µm)											
	0.4	0.5	0.6	0.7	0.8	0.9	1.0	1.1	1.2	1.3	1.4	1.5
andesite	10	15	19	22	23	23	24	25	26	27		
basalt	4	8	10	12	14	15	15	16	16	16		
altered basalt	28	36	50	59	61	62	62	63	66	66	60	60
altered basalt (w)	12	16	23	32	34	34	35	36	38	40	38	38
haematite	5	7	10	13	18	16	22	27	27	26	25	24
kaolinite	75	80	82	85	86	85	85	87	87	83	62	88
limestone	15	23	30	36	39	40	41	42	45	47		
montmorillonite	28	33	40	44	45	45	45	48	55	60	55	62
quartz	88	89	90	92	92	93	94	95	96	97	98	99
rhyolite	33	41	52	62	63	64	66	70	73	72	66	67
rhyolite (w)	15	20	28	37	40	43	46	48	50	52	48	50
shale	8	11	15	18	20	22	25	28	30	33		
tuff	36	39	43	57	60	60	59	59	57	56	55	55
tuff (w)	10	12	22	32	35	35	37	40	42	42	41	42

Visible-band average albedos (per cent)

granite	19
limestone	22
dolomite	15

Thermal infrared

Typical values of *emissivity* in the 8–12 µm band, unless otherwise stated.

basalt, polished	0.91
basalt, rough	0.94
dolomite, rough	0.96
dolomite, polished	0.93
feldspar, polished	0.82
feldspar	0.87
granite, polished	0.80
granite, rough	0.89
marble, polished	0.94
obsidian	0.86
quartz, polished	0.68
sandstone, polished	0.91

Microwave region

At radar wavelengths, dry rock has similar dielectric properties to dry *soil*, i.e. $\varepsilon \approx 3$. The dielectric constant increases roughly linearly with increasing moisture content.

ROSGYDROMET (Russian Federal Service for Hydrometeorology and Environmental Monitoring) See *CEOS*.

RSA (Russian Space Agency) See *CEOS*.

Run-length coding A technique for the reversible compression of digital image data. The image is considered as a one-dimensional array of digital numbers (e.g. by reading it from left to right and from top to bottom). It is encoded by recording the value of the first element in the array, and the number of subsequent contiguous pixels having the same value. Whenever the value changes, the new value is recorded, together with the length of the following 'run' of data which all have the same value.

Run-length coding is best suited to data that have a high degree of spatial homogeneity, for example classified images and smoothed images. In noisy images it is likely that the pixel value will change from one pixel to the next, so that most 'runs' of data will be only one pixel in length.

Compare *tesseral addressing*.

Russian Federal Service for Hydrometeorology and Environmental Monitoring See *CEOS*.

Russian Space Agency See *CEOS*.

RVI (Ratio vegetation index) See *vegetation index*.

S

S190A *Photographic system* carried on *Skylab* as part of the *EREP* package. Focal length: 152 mm. Wavebands: 0.40–0.90 μm (six bands, provided by six cameras). Film format: 57 × 57 mm. Image size: 163 × 163 km. Spatial resolution: 100 m.

S190B *Photographic system* carried on *Skylab* as part of the *EREP* package. Focal length: 457 mm. Waveband: 0.40–0.88 μm. Film format: 114 × 114 mm. Image size: 109 × 109 km. Spatial resolution: 55 m.

S191 U.S. optical/near infrared radiometer carried on *Skylab* as part of the *EREP* package. Wavebands: 0.4–2.4 μm, 6.2–15.5 μm. Spatial resolution: 450 m.

S192 U.S. optical/infrared imaging radiometer carried on *Skylab* as part of the *EREP* package. Wavebands: 13 channels between 0.4 and 12.5 μm. Spatial resolution: 79 m. Swath width: 72 km.

S193 Combined passive microwave radiometer, scatterometer and radar altimeter carried on *Skylab* as part of the *EREP* package. Frequency: K_u band (13.9 GHz). Spatial resolution: 16 km. Swath width: 180 km.

S194 Nadir-viewing passive microwave radiometer for soil moisture observations, carried on *Skylab* as part of the *EREP* package. Frequency: L band (1.4 GHz). Spatial resolution: 115 km.

SAGE I (Stratospheric Aerosol and Gas Experiment) U.S. limb-sounding (Sun-viewing) ultraviolet/optical/near infrared radiometer, for vertical profiling of aerosols, O_3 and NO_2 in the stratosphere, carried on *AEM-2*. Wavebands: 0.39, 0.45, 0.60, 1.0 μm.
The SAGE instrument was based on the *SAM-II* instrument.

SAGE II (Stratospheric Aerosol and Gas Experiment) U.S. limb-scanning optical/near infrared grating spectrometer, for vertical profiling of aerosols, CO_2, NO_2 and other trace gases in the atmosphere, carried on *ERBS*. Wavebands: seven channels between 0.385 μm and 1.02 μm.

SAGE III (Stratospheric Aerosol and Gas Experiment) U.S. limb-scanning ultraviolet/optical/near infrared spectrometer, for vertical profiling of atmospheric temperature, aerosols, O_3, CO_2, NO_2 and other trace gases, planned for inclusion on *Meteor-3M-1* and *ISSA* satellites as part of the *EOS* programme. SAGE III will view the Sun and Moon at occultation, using a diffraction grating to provide spectral resolution. Wavebands: nine channels between 0.29 and 1.55 μm. Height resolution: 1–2 km.

Salinity Ocean surface salinity can be measured by low-frequency (below about 5 GHz) passive microwave radiometry, since at these frequencies the *emissivity* is dominated by the effect of ionic conductivity (see *sea water*). The technique is accurate to about 1‰ if the effect of the *sea surface temperature* is allowed for, but the low frequencies involved mean that the technique has a very poor *spatial resolution* from a spaceborne platform.

Salyut Series of manned Soviet space stations operating between 1971 and 1989. Objectives: Earth remote sensing, zero-gravity experiments, astronomy. Orbit: Nominally circular *LEO* at (typically) 200 to 300 km altitude (Salyut-6, launched in 1977, was raised to about 500 km towards the end of its mission); inclination 52° (83° for Salyut 6). Principal instruments: *KATE-140*, *MKF-6*.

SAM II (Stratospheric Aerosol Measurement) U.S. Sun-viewing near infrared limb sounder, for measurement of stratospheric aerosol concentrations, carried on *Nimbus-7*. Waveband: 0.98–1.02 μm. Observation range: 5–40 km. Height resolution: 1 km.

SAMS (Stratosphere and Mesosphere Sounder) U.S. infrared radiometer, for vertical profiling of atmospheric temperature, H_2O, CH_4, CO and NO, carried on *Nimbus-7*. SAMS measured backscattered solar radiation. Waveband: nine channels between 4.1 and 15 μm and between 25 and 100 μm (gas cell modulation). Observation range: Up to 90 km. Height resolution: 10 km.

SAMS was operational until April 1985. An improved version, *ISAMS*, is carried on the *UARS* satellite.

SAP (Sensor AVP Package) Instrument consisting of two mechanically coupled optical/infrared imaging radiometers, carried by *DMSP* Block 5B/C satellites. Swath width: 3025 km.

High-resolution mode: Wavebands: 0.4–1.1 μm, 8–13 μm. Spatial resolution: 3.7 km (optical/near infrared), 4.4 km (thermal infrared).

Very high resolution mode: Wavebands: 0.4–1.1 μm, 8–13 μm. Spatial resolution: 620 m.

SAR
1. Acronym for *synthetic aperture radar*.
2. Generic name for synthetic aperture radar instruments carried on various satellites:

Almaz-1: Frequency: S band (3.13 GHz). Incidence angle: 30° to 60°. Spatial resolution: 15 m. Swath width: 40 km (spacecraft can roll to give coverage up to 350 km from nadir). Radiometric resolution: 3 dB.

JERS-1: Frequency: L band (1.28 GHz). Polarisation: HH. Incidence angle: 35° at mid swath. Spatial resolution: 18 m (3-look data). Swath width: 75 km.

Kosmos 1870: Frequency: S band (3.13 GHz). Spatial resolution: 25 m. Swath width: 20 km. Two instruments, looking left and right of sub-satellite track (spacecraft can roll to give coverage up to 250 km from nadir).

Radarsat: Frequency: C band (5.3 GHz). Polarisation: HH. Look direction: Right.

This is a multi-mode instrument. The different modes are summarised below:

Mode	Swath width (km)	Incidence angle (degrees)	Spatial resolution (m)		No. of looks
			Range	Azimuth	
standard	100	20–49	25	28	4
wide 1	165	20–31	30–48	28	4
wide 2	150	31–39	32–45	28	4
fine	45	37–48	10	9	1
scanSAR N	305	20–40	50	50	2–4
scanSAR W	510	20–49	100	100	4–8
scanSAR H	75	50–60	20	28	4
scanSAR L	170	10–23	28–63	28	4

Standard, fine and scanSAR H modes are programmable within the specified range of incidence angles.

Resurs-O2: Frequency: L band (1.3 GHz). Incidence angle: 35°. Spatial resolution: 100 m in range direction, 50 m in azimuth direction (when data are processed on the ground; 150 m when processed on board). Swath width: 50 km.

Seasat: Frequency: L band (1.3 GHz). Polarisation: HH. Incidence angle: 20°. Spatial resolution: 25 m (4-look data); pixel size 12.5 m. Swath width: 100 km. This was the first satellite-borne synthetic aperture radar.

SIR-A: Frequency: L band (1.3 GHz). Polarisation: HH. Incidence angle: 38°. Spatial resolution: 40 m. Swath width: 50 km.

SIR-B: Frequency: L band (1.3 GHz). Polarisation: HH. Incidence angle: 15–60°. Spatial resolution: 25 m. Swath width: 30–60 km.

SIR-C (L band): Frequency: 1.25 GHz. Polarisation: HH, VV, HV, VH. Incidence angle: 15–55°. Spatial resolution: 40 m. Swath width: 40–90 km.

SIR-C (C band): Frequency: 5.3 GHz. Polarisation: HH, VV, HV, VH. Incidence angle: 20–55°. Spatial resolution: 25 m. Swath width: 40–90 km.

SAR-3/SLR-3 Russian *synthetic aperture radar/side-looking radar* instrument carried on *Almaz-1B* satellite. Frequency: X band (8.6 GHz). Polarisation:

VV. Incidence angle: 38°–60° (SLAR mode), 25°–51° (SAR mode). Spatial resolution: SLAR mode: 190—250 m (range), 1200—2000 m (azimuth); SAR mode: 5—7 m. Swath width: 450 km (SLAR mode), 20–35 km (SAR mode); both to the left of the sub-satellite track. SAR swath can be adjusted between 190 and 500 km from the sub-satellite track by rolling the spacecraft.

SAR-10 Russian *SAR* instrument carried on *Almaz-1B* satellite. Frequency: S band (3.13 GHz). Incidence angle: 21°–51°. Spatial resolution: 15–40 m (wide swath), 15 m (intermediate swath), 5–7 m (narrow swath). Swath width: 120–170 km (wide), 60–70 km (intermediate), 30–55 km (narrow); both to the left and to the right of the sub-satellite track. Swaths can be selected within a band between 190 and 520 km from the sub-satellite track by rolling the spacecraft.

SAR-70 Russian *SAR* instrument carried on *Almaz-1B* satellite. Frequency: P band (430 MHz). Incidence angle: 25°–51°. Spatial resolution: 22–40 m. Swath width: 120–170 km, selected between 190 and 520 km to the left of the sub-satellite track by rolling the spacecraft.

SASS (Seasat-A Scatterometer System) U.S. microwave wind *scatterometer* carried on *Seasat*. Frequency: K_u band (15.0 GHz). Beam angles: [0° = forward, 90° = right] ±45°, ±135°. Incidence angle: 20° to 65°. Spatial resolution: 50 × 100 km. Swath width: 500 km. Near edges 200 km from sub-satellite track; far edges 700 km. Left and right. Accuracy: ±2 m/s (4 to 26 m/s); ±20° for ocean surface wind vectors.

Satelite de Sensoriamento Remoto See *SSR-1*.

Satellite Infrared Spectrometer See *SIRS*.

Satellite Pour L'Observation de la Terre See *SPOT*.

Saturation

1. A detector is said to be saturated when any further increase in the input signal will not result in an increase in the output signal. This phenomenon can be caused either by non-linearity in the detector itself or, more commonly in the context of remote sensing instruments, by the fact that the process of digitising the detected signal imposes a maximum measurable value. For example, a detector whose output is digitised into 8-bit numbers can record 256 different signal values. If the output is proportional to the input, this means that the system will become saturated at a signal 255 times as large as the minimum measurable signal. Thus the tendency to saturate can only be reduced at the expense of decreasing the *radiometric resolution*, or by increasing the number of bits in the digitised output.
2. See *IHS display*.

SAVI (Soil-adjusted vegetation index) See *vegetation index*.

S band Subdivision of the microwave region of the *electromagnetic spectrum*, covering the frequency range 1.55 to 5.2 GHz (wavelengths 58 to 193 mm).

SBUV (Solar Backscatter Ultraviolet Radiometer) U.S. nadir-viewing ultraviolet filter spectrometer, for measuring vertical ozone profiles, carried on *Nimbus-7* and *NOAA-9* onwards. SBUV, which is a development of *BUV*, measures vertically backscattered radiation, and also views the Sun directly for calibration. Waveband: 12 channels between 0.16 and 0.40 µm. Spatial resolution: 200 km (Nimbus), 170 km (NOAA).

Scalar approximation See *Kirchhoff model*.

Scale For maps and, by extension, for remotely sensed images of the Earth's surface, the scale is defined as the ratio of the length (linear dimension) of the representation of an object in the map or image to the real length of the object. **Large-scale** is generally taken to mean having a scale of greater than 1/50 000 and **small-scale** less than 1/500 000.

The scale of an image can be increased by enlargement. However, the coarsest desirable *spatial resolution* on a map is about 0.5 mm, and this imposes an upper limit, set by the spatial resolution of the imagery, on the scale of a map that can be constructed from the image. Large-scale mapping therefore requires a spatial resolution of 25 m or better; small-scale mapping requires a resolution of about 250 m.

SCAMS (Scanning Microwave Spectrometer) U.S. scanning passive microwave radiometer, carried on *Nimbus-6*. Frequencies: 22.2, 31.7, 52.9, 53.9, 55.5 GHz. Spatial resolution: 145 km at nadir. Swath width: 2600 km. Sensitivity: 0.2 K to 0.6 K. Absolute accuracy: 1.5 K.

Scanner for Earth Radiation Budget See *Scarab*.

Scanning Normally, the process by which a remote sensing system having limited instantaneous coverage generates an *image* of the Earth's surface.

In the context of *electro-optical* systems for the detection of **optical and infrared** radiation, a scanner is a remote sensing system which uses mechanical scanning to generate images of the Earth's surface. It possesses three advantages over *photographic systems*: very fine radiometric resolution in narrow and simultaneously recorded wavebands; *detectors* that can collectively span a relatively large range of electromagnetic radiation; data that can be stored in digital form for digital *image processing*.

A multispectral scanner measures the radiation from the Earth's surface along a scan line perpendicular to the line of the aircraft or satellite path. As the platform moves forward, repeated measurement of the radiation allows a two-dimensional image to be built up. These scanners comprise a collecting section, a detecting section and a recording section. A telescope directs radiation onto a rotating mirror, which reflects the radiation into the instrument's

189

optical system, providing a narrowly focussed beam. This beam may be split, usually by a dichroic (doubly refracting) element, into its reflected (optical and near infrared) and emitted (thermal infrared) components. The reflected component is spectrally dispersed, using a set of filters, a prism or a diffraction grating, and focussed on suitable *detectors*. The optical and near infrared radiation is usually detected by silicon photodiodes. The thermal infrared radiation is usually detected by photon detectors, held in a cooling flask.

After detection, the signals are amplified and formatted for recording or direct transmission, either by being fed into a film recorder to produce images directly, or by digitisation for later image processing. Current scanners produce a digital output which is recorded by a digital tape recorder or transmitted to a ground receiving station.

Many airborne and satellite sensors are scanners of this type, although these are now being superseded by sensors using *charge-coupled devices* (CCDs), either wholly or in conjunction with a simplified scanning system. See also *spin-scan imager, step-stare imager, vidicon*.

Scanning of a **microwave** *antenna* in angle may be accomplished through mechanical means, by physically rotating the antenna or a reflecting surface. Alternatively, the beam can be made to scan electronically through the use of an antenna array with electrically controlled phase shifters. Electronic scanning requires no physical movement of the antenna, and the beam can be steered at very high rates compared with mechanical scanning, but electronically scanned antennas are heavier, more expensive to build, and the presence of the phase shifters degrades the radiation efficiency (see *antenna temperature*).

Scanning Imaging Absorption Spectrometer for Atmospheric Cartography See *Sciamachy*.

Scanning Microwave Spectrometer See *SCAMS*.

Scanning Multichannel Microwave Radiometer See *SMMR*.

Scanning Radiometer
1. See *SR*.
2. Chinese optical/infrared spin-scan imaging radiometer, carried on *FY-2* satellite. Wavebands: 0.55–1.05 μm (4 bands), 6.2–7.6 μm, 10.5–12.5 μm. Spatial resolution: 1.3 km at nadir (optical/near infrared), 5 km at nadir (mid- and thermal infrared). Field of view: Full Earth disc as seen from *geostationary* orbit. Scan time: 30 minutes (full disc).

Scanning telephotometer Instrument carried by *Meteor* satellites, operating in the waveband 0.5 μm to 7.0 μm and giving a spatial resolution of 2 km (global coverage) or 1 km (local coverage).

Scanning TV radiometer See *STR*.

Scarab (Scanner for Earth Radiation Budget) French–German–Russian radiometer for Earth radiation budget measurements, to be included on *Meteor-3-7*, *Envisat* and *Metop*. Wavebands: 0.2–50, 0.2–4, 0.5–0.7, 10.5–12.5 µm. Spatial resolution: 60 km. Swath width: 3000 km.

URL: http://www.cnes.fr/Etude_gest_plan/scarab/index.html

Scattering Deflection of electromagnetic radiation, without absorption, as a result of its interaction with particles (electrons, atoms, molecules or macroscopic particles such as water droplets) or a solid or liquid surface. See *backscatter, Mie scattering, radiative transfer equation, Rayleigh scattering*.

Scattering models Scattering models are mathematical expressions relating radiation quantities resulting from a wave incident on a scene (target) to the physical and geometric parameters of the scene. The development of these models may be based on physical laws, intuitive understanding of the phenomena, or empirical data fitting. Scattering models based on physical principles are usually more complex than other types of models, but they normally have a wider range of applicability. An empirical model is usually the simplest, but it may not be possible to extrapolate it to other data sets or to obtain from it an understanding of the effect of the scene parameters. A model based on an intuitive understanding of a physical phenomenon is intermediate in complexity and usually accounts only for a given scattering mechanism believed to be dominant for a given scene and under a given system or geometry.

Depending on whether the reradiation comes from a surface area or throughout the volume of a medium, models are characterised as surface or volume scattering models. Whether a scene should be modelled as a surface or a volume depends upon the geometric and physical properties of the scene as well as the incident wavelength. For example, the reradiation from a soil medium is usually represented by a surface scattering model at microwave frequencies. This is because the soil particles are of the order of tens to hundreds of microns across, while the incident wavelength is of the order of centimetres. That is, the effect of inhomogeneity is small compared with the average dielectric discontinuity at the air–soil interface. On the other hand, at visible wavelengths the scattering by the same soil medium must be represented by a volume scattering model because the incident radiation senses a discrete collection of particles rather than a continuous soil surface.

See *Bragg scattering, dense medium models, facet model, integral equation model, Kirchhoff model, radiative transfer models, small perturbation model.*

Scatterometer Active microwave remote sensing instrument, designed to provide accurate measurement of the surface backscatter coefficient, usually as a function of look direction. Scatterometers differ from imaging *radars* in giving higher precision (radiometric resolution) at the expense of spatial resolution.

Their major application is in determining *wind velocity* over ocean surfaces. The backscatter coefficient σ^0 (in dB) can be described by the expression

$$\sigma^0 = aU^\gamma(1 + b\cos\psi + c\cos 2\psi),$$

where U is the wind speed, ψ is the angle between the wind vector and the horizontal component of the scatterometer look direction, and a, b, c and γ are constants for a given frequency, incidence angle and polarisation state. The dependence on ψ allows the wind direction to be determined from scatterometer observations made in several directions; in particular, the term in 2ψ allows the upwind/downwind ambiguity to be resolved, although the coefficient c is generally significantly smaller than b.

Scatterometer data are also increasingly being used to study *sea ice* distributions and motion vectors, and have also been applied to monitoring of rain forests, tundra and deserts.

The table below summarises the principal spaceborne scatterometers. The horizontal components of the look directions are specified such that $0°$ is forward and $90°$ to the right of the sub-satellite track.

Instrument	Satellite(s)	Frequency (GHz)	Polarisation	Spatial res. (km)	Swath width (km)	Look directions (degrees)
AMI-SCAT	ERS-1, -2	5.3	VV	50	500	45, 90, 135
ASCAT	Metop	5.3	VV	50	500×2	-135, -90, -45, 45, 90, 135
SASS	Seasat	14.6	HH, VV	50	500×2	-135, -45, 45, 135
NSCAT	ADEOS	14.0	HH, VV	25	600×2	-135, -65, -45, 45, 115, 135
SeaWinds	ADEOS-2	14.0	HH, VV	50	600×2	conical scan

Sciamachy (Scanning Imaging Absorption Spectrometer for Atmospheric Cartography) German/Dutch instrument for profiling tropospheric and stratospheric trace gases, planned for inclusion on *Envisat*. Sciamachy is a differential absorption ultraviolet/optical/infrared spectrometer that can operate as a limb sounder or as a nadir sounder. Wavebands: 0.24–1.70 μm, 1.98–2.02 μm, 2.27–2.38 μm. Spatial resolution: 32×320 km at nadir, 3 km height resolution as limb sounder.

URL: http://envisat.estec.esa.nl/instruments/sciamachy/index.html

SCMR (Surface Composition Mapping Radiometer) U.S. optical/infrared imaging radiometer, carried on *Nimbus-5* satellite. Wavebands: 0.8–1.0, 8.3–9.3, 10.5–11.3 μm. Spatial resolution: 800 m. Swath width: 800 km.

The instrument failed a few weeks after launch in 1972.

SCR (Selective Chopper Radiometer) U.S. nadir-viewing infrared radiometer for atmospheric profiling, carried on *Nimbus-4* and *Nimbus-5* satellites. Waveband: six channels near 15 μm (Nimbus-4); 16 channels from 2.1 μm to 133 μm (Nimbus-5). Spatial resolution: 25 km.

Sea ice Sea ice is ice formed by the freezing of sea water, at a temperature of $-1.9\,°C$ or less, predominantly in the Arctic and Southern Oceans. It is of great significance to the global climate system through its effect on the global *albedo*, the role it plays in determining the lower boundary temperature of the atmosphere, and the physical restriction it imposes on the transport of heat, water, gases and momentum between the atmosphere and the ocean. Knowledge of sea ice distribution is also required by shipping and offshore activities.

Sea ice undergoes significant physical changes over time, and as a consequence many different forms of ice are recognised. The most important distinction is between **first-year ice**, which is less than 1 year old, and **multi-year ice**, which has survived at least one melt season. Sea ice is a dynamic material, moved by wind and ocean stresses, and as the ice ages it tends to become thicker and more irregular in profile as a result of the collision and coalescence of individual floes. Sea ice rejects salts during freezing and as it ages. This alters the salinity and density of the upper ocean, and can lead to deep convection and the formation of bottom water. Sea ice also represents a significant navigational hazard.

Local investigations of sea ice can be made using visible-wavelength or high-resolution *synthetic aperture radar* (SAR) remote sensing imagery, in which individual ice floes (of the order of 30–300 m in size) can often be resolved. *Landsat MSS* and *TM* imagery, and *SPOT HRV* imagery etc., provide the high spatial resolution necessary to resolve the smaller floes, but these instruments are somewhat limited temporally since the *Sun-synchronous orbits* of their satellites mean that, poleward of latitude 70°, the level of solar illumination is adequate only between late October and late March (Antarctic) and between late April and late September (Arctic). The analysis of SAR imagery shows particular promise as a technique for identifying ice type (through image *texture*, and through the analysis of polarimetric SAR data), and its insensitivity to cloud cover means that the time-series of images necessary for the investigation of ice **dynamics** (by tracking individual floes or by image correlation or wavelet analysis techniques) can often be obtained. Ice dynamics at larger, synoptic, scales can also be obtained from analysis of *passive microwave* and *scatterometer* data.

The spatial coverage of visible-wavelength and SAR imagery that is capable of resolving most floes is inadequate for synoptic studies of sea ice distribution. For this purpose, passive microwave data are particularly valuable, since the *emissivity* of sea ice is significantly higher than that of open water. Scatterometer data are also finding increasing application to ice concentration measurements. The spatial resolution of these techniques is inadequate to resolve individual floes (except for very large floes, or for observations made from airborne rather than spaceborne platforms), so the technique is used to estimate the ice **concentration**, i.e. the fractional area represented by a particular ice type. Multifrequency observations can be used to discriminate a number of different ice types, although the most widely used algorithms only determine the concentrations of first-year and multi-year ice. For example, the four-channel algorithm due to Cavalieri *et al.* used brightness temperatures from the *SMMR* instrument at horizontal and vertical polarisations and at 18 and 37 GHz. A polarisation ratio P and a gradient ratio G were defined as follows:

$$P = \frac{T_{V18} - T_{H18}}{T_{V18} + T_{H18}},$$

$$G = \frac{T_{V37} - T_{V18}}{T_{V37} + T_{V18}},$$

and the concentrations C_i of first-year ($i = 1$) and multi-year ($i = 2$) ice calculated from expressions of the form

$$C_i = \frac{a_i + b_i P + c_i G + d_i PG}{e_i + f_i P + g_i G + h_i PG}.$$

The values of P and G corresponding to open water are 0.263 and 0.083 respectively; those for Arctic first-year ice are 0.036 and -0.005, and those for Arctic multi-year ice 0.059 and -0.075. This algorithm is generally accurate to about 5%. However, it is not valid when the ice surface is wet (i.e. in summer, or near the ice edge). Slightly different coefficients are used in the Antarctic, where the multi-year concentration is in any case not generally calculated since it is small.

Sea ice, electromagnetic properties (Microwave region) The *dielectric constant* of sea ice is strongly dependent on frequency, temperature, and the composition (notably the salinity) of the ice. At temperatures below about $-5\,^\circ\text{C}$, the dielectric constant is described reasonably accurately by linear models:

$$\varepsilon' = A + Bv,$$

$$\varepsilon'' = C + Dv,$$

where v is the volume fraction of brine. Appropriate values of A, B, C and D are given below:

f (GHz)	A	B	C	D
1	3.12	0.009	0.04	0.005
4	3.05	0.0072	0.02	0.0033
10	3.0	0.012	0.00	0.010

At higher temperatures, both ε' and ε'' are larger.

Emissivities of dry first-year and multi-year ice vary with frequency roughly as shown below:

f (GHz)	First-year	Multi-year
5	0.93	0.93
10	0.92	0.89
20	0.94	0.83
50	0.95	0.73
100	0.95	0.64

See also *ice*.

Seasat U.S. satellite, operated by *NASA*, launched in June 1978. Mission terminated after 3 months. Objectives: Experimental environmental satellite, primarily intended for oceanography but with important applications to land-ice mapping. Orbit: Circular *LEO* at 800 km altitude. Period 800 km; inclination 108°. Principal instruments: *Radar Altimeter, SAR, SASS, SMMR*.
Seasat was the first satellite to carry an imaging radar.

URL: http://www.jpl.nasa.gov/mip/seasat.html

Seasat-A Satellite Scatterometer See *SASS*.

SeaStar U.S. satellite launched in August 1997. The nominal lifetime is 5 years. Objectives: Dedicated ocean colour mission. Orbit: Circular *Sun-synchronous LEO* at 705 km altitude. Period 98.9 minutes; inclination 98.2°; equator crossing time 12:00 (descending node). *Exactly repeating orbit* (233 orbits in 16 days). Principal instruments: *SeaWiFS*.
SeaStar is also known as Orbview-2, and is operated by *Orbimage*.

URL: http://seawifs.gsfc.nasa.gov/SEAWIFS/SEASTAR/
SPACECRAFT.html

Sea-state bias See *electromagnetic bias*.

Sea surface temperature The Sea Surface Temperature (SST) is an important parameter in determining the lower boundary condition for the temperature of the atmosphere over oceans, and hence in meteorological and climate modelling. The global SST distribution is used to initialise and validate global circulation models and to study global heat exchange between ocean and atmosphere. At smaller physical scales, observations of the SST can reveal ocean circulation features such as eddies, fronts and upwellings, and can also be used to identify areas likely to have high biological productivity.

SST can be determined from satellite remote sensing methods using either *thermal infrared* or *passive microwave radiometry*. Thermal infrared observations are made at wavelengths near 3.7 μm or near 10 μm. The 3.7 μm band is more sensitive, but is subject to contamination by reflected sunlight, and so is used only for night-time observations. Observations at this wavelength also require correction for the effect of *aerosols*. Thermal infrared measurements in either waveband require atmospheric correction, and the detection and removal of pixels contaminated by cloud.

Although they provide a lower spatial resolution than thermal infrared measurements, passive microwave observations of SST are less significantly affected by cloud and are easier to correct for atmospheric effects. Errors are introduced by wind-generated roughness and by *precipitation*, but these can mostly be removed by suitable multifrequency observations. A typical passive microwave instrument designed for SST measurements might have channels at 7, 11, 18, 21 and 37 GHz. The 7 GHz band provides the 'raw' measurement and the 21 GHz band provides the major atmospheric correction, for *water vapour*. The remaining bands are used to correct for the effects of surface roughness, clouds etc.

195

Both thermal infrared and passive microwave remote sensing of SST can be subject to a systematic error. The temperature inferred from these measurements is that of the upper surface of the water, of the order of the *attenuation length*. The oceanographically important parameter is usually the mean temperature of the upper few centimetres of the water column, and this can differ from the skin temperature by an amount of the order of 1 K as a result of diurnal heating and evaporative cooling effects. For accurate determination of the SST, these effects must be modelled and corrected for.

Sea surface topography The mean surface of the sea is generally within ± 100 m of the *ellipsoid*, and within ± 1 m of the *geoid*, once the effects of ocean currents and of time-varying effects (tides, surface gravity waves, wind-driven variations and atmospheric pressure effects) have been removed by modelling or by time-averaging. Deviations from this mean surface provide information on ocean **circulation** (e.g. mesoscale [50–500 km] features such as eddies, large-scale seasonal changes, thermal expansion effects and inter-annual variations such as the El Niño oscillation). Ocean circulation in turn provides information on oceanic heat transport and on the transport of nutrients and dissolved gases. Sea surface topography also provides information on sea-floor topography.

Sea surface topography can be determined from satellite measurements using *radar altimeter* observations, with an accuracy of better than a few cm.

See also *electromagnetic bias, ocean currents and fronts, topographic mapping.*

Sea-Viewing Wide Field Sensor See *SeaWiFS*.

Sea water, electromagnetic properties
Optical/infrared region
For the optical and near infrared region, see *ocean colour*.
Typical thermal infrared *emissivity* (10–12 μm): 0.99.

Microwave region
The real part of the dielectric constant of sea water in the microwave region can be described by the *Debye equation*, but the imaginary part is modified to take account of the conductivity σ of the water:

$$\varepsilon'' = \frac{\omega \tau \varepsilon_p}{1 + \omega^2 \tau^2} + \frac{\sigma}{\varepsilon_0 \omega}.$$

ε_∞ can be taken as 5.0. The table shows the values of ε_p, τ and σ for a salinity of 35‰.

T (°C)	ε_p	τ (ps)	$\sigma\ (\Omega^{-1}\,m^{-1})$
0	72.8	17.0	2.9
10	70.6	12.3	3.8
20	67.5	9.1	4.8
30	64.7	7.1	6.0
40	63.6	5.8	7.4

See also *water.*

SeaWiFS (Sea-viewing wide field of view sensor) U.S. optical/near infrared imaging radiometer, carried on *SeaStar* satellite. Wavebands: 402–422, 433–453, 480–500, 500–520, 545–565, 660–680, 745–785, 845–885 nm. Spatial resolution: 1.1 km at nadir (4.5 km for global area coverage). Swath width: 1500 km, 2800 km (global area coverage).

Ocean colour instrument. SeaWiFS was originally intended to fly on *Landsat-6*.

URL: http://seawifs.gsfc.nasa.gov/SEAWIFS.html

SeaWinds U.S. conically scanned microwave wind scatterometer, planned for inclusion on *ADEOS-II*. Frequency: K_u band. Incidence angles: 40°, 46°. Spatial resolution: 50 km. Swath width: Two 600 km swaths (left and right). SeaWinds is derived from *NSCAT*. It is also known as NSCAT-II.

URL: http://winds.jpl.nasa.gov/

Segmentation An operation in image processing, in which contiguous homogeneous regions are identified within the image. Homogeneity is generally defined after *classification* of the image (for example, a simple *density slicing* operation), so that a homogeneous region is one in which all the pixels belong to the same class. Contiguity is then defined by identifying the boundary of the region, often by using an *edge detection* algorithm, and performing morphological operations on the boundary to ensure that it forms a closed loop with no gaps or dangling ends.

Selective Chopper Radiometer See *SCR*.

Semimajor axis See *orbit, satellite*.

Semiminor axis See *orbit, satellite*.

Semivariogram A simple and reasonably intuitive method of describing the spatial variation of some property P, for example the digital numbers of the pixels in an image. The semivariogram is the function

$$\gamma(\delta) = \tfrac{1}{2}\langle[P(\mathbf{x} + \delta) - P(\mathbf{x})]^2\rangle,$$

where $P(\mathbf{x})$ is the value of P at some position $\mathbf{x}$, δ is a spatial separation, and $\langle\ \rangle$ denotes a spatial average. For a statistically stationary property P, γ is related to the autocorrelation function ρ by

$$\gamma(\delta) = \sigma^2[1 - \rho(\delta)],$$

where σ is the variance of P. However, the semivariogram can be defined even for properties that are not statistically stationary and hence do not have well-defined autocorrelation functions.

The semivariogram is sometimes called the structure function. It has found some application in the characterisation of image *texture*.

Sensitivity See *radiometric resolution.*

Sensor In remote sensing, the instrument that collects electromagnetic radiation and converts it to some other form, usually a digitised electronic signal.

Sensor AVP Package See *SAP.*

Separability A measure of the extent to which two probability distributions do not overlap, and hence of the extent to which two clusters in feature space (see *clustering*) can be used in the *classification* of an image to represent different image classes. If the two probability distributions are $p_1(\mathbf{x})$ and $p_2(\mathbf{x})$ respectively, where $\mathbf{x}$ represents a position vector in feature space, the **divergence** d is defined as

$$d = \int (p_1(\mathbf{x}) - p_2(\mathbf{x})) \ln \frac{p_1(\mathbf{x})}{p_2(\mathbf{x})} \, d\mathbf{x},$$

where the integral is carried out over all feature space. The divergence is zero for identical distributions, and infinite for non-overlapping distributions. The **Bhattacharyya distance** B is defined in terms of the mean values $\langle \mathbf{x}_1 \rangle$ and $\langle \mathbf{x}_2 \rangle$ and the *covariance matrices* $\mathbf{C}_1$ and $\mathbf{C}_2$ as

$$B = \frac{1}{4}(\langle \mathbf{x}_1 \rangle - \langle \mathbf{x}_2 \rangle)^T (\mathbf{C}_1 + \mathbf{C}_2)^{-1}(\langle \mathbf{x}_1 \rangle - \langle \mathbf{x}_2 \rangle) + \frac{1}{2} \ln \frac{|\mathbf{C}_1 + \mathbf{C}_2|}{2\sqrt{|\mathbf{C}_1||\mathbf{C}_2|}},$$

where the superscript T denotes a vector transpose. Like the divergence, this is zero for identical distributions and infinite for non-overlapping distributions. Separability measures that tend to finite values for non-overlapping distributions include the **Jeffries–Matusita distance**

$$J = \int \left(\sqrt{p_1(\mathbf{x})} - \sqrt{p_2(\mathbf{x})} \right)^2 d\mathbf{x}$$

and the **transformed divergence**

$$d^* = 2(1 - \exp[-d/8]),$$

both of which take a value of 2 for non-overlapping distributions. For normally-distributed multivariate data, J and B are related by

$$J = 2(1 - \exp[-B]).$$

The probability that the correct choice will be made in classifying a pixel into one of two classes having a divergence d and equal prior probabilities has a maximum possible value of

$$1 - \frac{\exp(-d/2)}{8}.$$

As a rough guide for real data, the probability is 0.5 for $d \approx 2$ and 0.9 for $d \approx 20$.

SEVIRI (Spinning Enhanced Visible and Infrared Imager) European multiband spin-scan optical and infrared imager planned for inclusion on *MSG* satellites. Wavebands: 0.51–0.73, 0.56–0.71, 0.71–0.95, 1.44–1.79, 3.4–4.2, 5.4–7.2, 6.9–7.9, 8.3–9.1, 9.5–9.9, 9.8–11.8, 11.0–13.0, 13.0–13.8 µm. Spatial resolution: 3 km (1 km for 0.51–0.73 µm channel). Field of view: Full Earth disc as seen from geostationary orbit. Scan time: 12 minutes (plus 3 minutes between scans).

SFOR-1 Russian ultraviolet spectrometer planned for inclusion on *Meteor-3M* satellites. Wavebands: eight channels between 0.279 and 0.380 µm. Spatial resolution: 25 km.

SFOR-1 is a backscatter radiometer for determining atmospheric ozone concentrations.

Shadowing, radar The prevention of electromagnetic radiation from a *side-looking radar* or *synthetic aperture radar* from reaching some part of the Earth's surface as a result of its interception by another part of the surface. Unlike optical shadows, which receive some *diffuse illumination* from the sky, no information at all is received from a radar shadow.

Shadow sites in archaeology These are visible archaeological sites where the remains are still partly above the surface. They are seen from the air in the early morning or late afternoon through the shadows cast by above-surface irregularities. Vegetation stabilises the soil over sites which have decayed. In desert or near-desert areas, many sites of this type survive. In temperate climates, height differences are small and the surface irregularity broad, and they cannot usually be seen from normal surface viewing distances. Ploughing and other modern large-scale earth-moving operations destroy the visible remains of such sites.

Contrast is a function of height of the remains, time, date and latitude of the photograph, angle of view and intrinsic surface colour. It is best when the photograph is taken opposite to the direction of the Sun's illumination. Rephotographing in different seasons is necessary to record complex details. In the spring and summer, contrast improves in the hour before sunset. In winter, best light is usually available after the passage of a cold front in the late afternoon. Height differences can be more easily seen stereoscopically with two pictures made at roughly the same angle in oblique views or in vertical images.

See also *archaeological site detection*.

Shape detection A high-level operation in image processing is the recognition and classification of objects on the basis of their geometry. Various approaches to shape detection are possible. Linear features can be detected using *edge detection* operators, unless they are only a single pixel in width, in which case the

Hough transform or a *line detection* operator can be used. Other approaches include template matching, where a *convolution operator* is defined with the appropriate shape, size and orientation to detect a particular feature, and the use of *segmentation* followed by characterisation of the shape of the segment. This characterisation can, again, be performed in a number of ways, for example by calculating the *Fourier transform* of the boundary or by calculating shape factors or moments of area.

Sharpening Sharpening of an image is the process of increasing the amplitude of the higher spatial frequencies while leaving the lower spatial frequencies unchanged. A sharpening operator is thus sometimes called a **high-boost filter**. The most obvious effect of sharpening an image is the enhancement of the visibility of edges between homogeneous regions, but at the expense of increasing the noise level.

Image sharpening is normally performed using a *convolution operator*. If S is a *smoothing* operator and I is the operator that leaves an image unchanged (the identity operator), the operator

$$k\mathbf{I} + (1 - k)\mathbf{S},$$

where $k > 1$ is the degree of sharpening, is a sharpening operator. For example, if S is a 3×3 average and $k = 2$, the operator has the following kernel:

$-1/9 \quad -1/9 \quad -1/9$

$-1/9 \quad 17/9 \quad -1/9$

$-1/9 \quad -1/9 \quad -1/9$

Another common sharpening operator is

$$\mathbf{I} + k\mathbf{L},$$

where $\mathbf{L}$ is the Laplacian operator (see *edge detection*).

Sheen, oil See *oil spills*.

SHF Russian scanning passive microwave radiometer, for *SST* and atmospheric temperature measurements, carried on *Meteor-Priroda* satellites. Frequencies: 7.5, 19.4, 22.2, 37.5 GHz. Spatial resolution: 30 km at 7.5 GHz to 13 km at 37.5 GHz. Swath width: 800 km.

Shuttle Imaging Radar See *SIR-A, -B, -C*.

Side lobe See *power pattern*.

Side-looking radar (SLR) An active microwave imaging technique, also known as side-looking airborne radar (SLAR) or real aperture radar (RAR), which produces a two-dimensional representation of the spatial variation of the radar *backscatter coefficient*. Figure A shows the geometry of SLR imaging, over a flat Earth for simplicity:

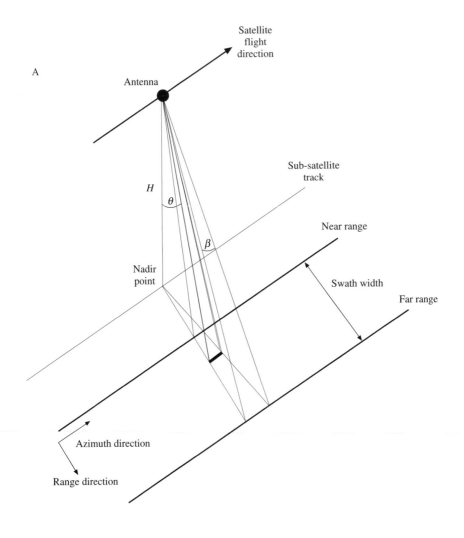

A

The antenna views to one side of nadir only, and transmits very short pulses of microwave radiation. At some instant of time, the area on the ground contributing to the signal received at the antenna as a result of the reflection of a particular pulse is the area shaded in the diagram. As time increases, the location of this region moves from near range to far range as a result of the increasing two-way travel time from the antenna to the reflecting region. The coordinate perpendicular to the sub-satellite track, corresponding to increasing delay in the travel time, is called the **range direction** or the **across-track direction**. The perpendicular coordinate is called the **azimuth direction** or the **along-track direction**.

The spatial resolution of the system is determined by the size of the region contributing to the signal received at some instant of time, i.e. the shaded area in the figure. The azimuth resolution R_a is governed by the antenna's *beamwidth* β in the azimuth direction, and can be written as

$$R_a = R\beta,$$

where R is the slant range, i.e. the distance from the antenna to the ground. Since

$$\beta \approx \lambda/L,$$

where λ is the wavelength and L is the length of the antenna measured in the flight (azimuth) direction, and since, over a flat Earth, the slant range is given by

$$R = H/\cos\theta,$$

where θ is the incidence angle and H is the height of the radar above the ground; we must have

$$R_a \approx \frac{H\lambda}{L\cos\theta}.$$

The range resolution R_r is governed by the duration t_p of the pulse, since two points having the same azimuth coordinate will be resolved only if their two-way travel times differ by at least t_p. Over a flat Earth, this gives a range resolution of

$$R_r = \frac{ct_p}{2\sin\theta},$$

where c is the speed of light.

The *swath width* of an SLR system is the distance between near and far range, defined by the antenna's beamwidth in the range direction. This cannot be increased independently of the radar's other imaging parameters, since a large swath width implies a large difference between the minimum and maximum two-way travel times. In order to ensure that there can be no ambiguity in identifying the transmitted pulse responsible for the signal detected at a given instant, this places an upper limit on the pulse repetition frequency, and hence on the sampling rate in the azimuth direction.

The side-looking geometry of the SLR, necessary to eliminate ambiguity in the pulse return time between the left and right sides of the sub-satellite track, gives rise to some important geometric effects. The parameter corresponding to the range coordinate that is actually measured by the radar is the slant range R, and SLR images are often presented in a slant range – azimuth projection. Since the slant range does not vary linearly with the ground range even over a flat surface, this introduces a geometric distortion. This **slant range distortion** can be corrected for flat surfaces if the imaging geometry is known; over varying topography this distortion results in **layover**, in which topographic features appear to lean towards the near-range edge of the image. This is due to the fact that a scatterer at a higher elevation will have a smaller slant range than one with the same range and azimuth coordinates but lower elevation, and will thus be assigned a lower range coordinate. Layover effects can be corrected if the topography is known. Slant range distortion effects increase with decreasing incidence angle. **Highlighting** occurs when topographic variations cause parts of the surface to be viewed at or near normal incidence, causing increased values of the backscatter coefficient. For example, the surface AB in figure B will appear much less bright than the surface BC. Variability in images due to this effect increases as a result of increasing incidence angle.

Shadowing is caused when one part of the surface prevents the radar signal from reaching another, as shown in figure B. This problem becomes more acute for larger (shallower) incidence angles.

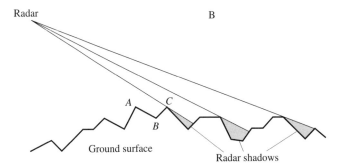

The range resolution of an SLR is independent of the height H of the radar, being determined only by the pulse length and the imaging geometry. The azimuth resolution, however, is proportional to H and can only be improved by increasing the length L of the antenna. For a spaceborne radar at $H = 800$ km operating at $\lambda = 30$ cm, the azimuth resolution is of the order of 25 km even for a 10-m long antenna. Since longer antennas are impracticable, higher resolution is achieved using the *synthetic aperture radar* technique. Although they are more commonly used on airborne platforms, from which high azimuth resolutions can be achieved by using lower values of H, SLRs have been placed in space, for example on the *Okean-O-1* and *Almaz-1B* satellites.

Sidereal day The period of the Earth's rotation about its axis, relative to the 'fixed stars'. It is equal to

$$\frac{y}{y+1}$$

solar days of 24 hours, where y is the number of solar days in the average year. Since $y \approx 365.242$, the sidereal day is approximately 86 164 seconds or 23 hours 56 minutes 4 seconds.

SI-GDR (Spectrometer/Interferometer) German nadir-viewing infrared Fourier spectrometer, for atmospheric temperature profiling, carried on the *Meteor-Priroda* series of satellites. Waveband: 6.25 to 25 μm. Spectral resolution: 0.6 MHz. Spatial resolution: 25 km.

Significant wave height The Significant Wave Height (SWH) is the simplest and most widely reported statistic describing sea state, and can be determined by *radar altimeter* observations. It is normally defined as the mean height (from trough to crest) of the highest third of the observed waves, and denoted by the symbol $H_{1/3}$. It is related to the variance σ^2 of the surface height distribution by $H_{1/3} = 4\sigma$. The SWH is determined by the wind speed, the **fetch** (distance from the shore, measured in the direction in which the wind is blowing) and the **duration** (time for which the wind has been blowing). For sufficiently large fetch and duration, the sea is said to be **fully developed**, in which case the SWH depends only on the wind speed and is given approximately by

$$\frac{H_{1/3}}{\text{m}} = 0.021 \left(\frac{U_{10}}{\text{m s}^{-1}}\right)^2,$$

where U_{10} is the wind speed 10 m above the surface.

Sinc function A mathematical function important in image and signal processing. The sinc function is usually defined as

$$\text{sinc}(x) = \frac{\sin x}{x},$$

so that it has zeros at $x = \pm\pi, \pm2\pi, \pm3\pi$ etc., and $\text{sinc}(0) = 1$. It is occasionally defined as

$$\text{sinc}(x) = \frac{\sin \pi x}{\pi x},$$

in which case the zeros are at $x = \pm1, \pm2, \pm3$ etc. The importance of the sinc function derives from the fact that it is the *Fourier transform* of a rectangular pulse function.

Single-frequency Solid State Radar Altimeter See *SSALT*.

Sinusoidal projection See *map projection*.

SIR-A, -B, -C (Shuttle Imaging Radar) Series of *Space Shuttle* missions carrying *synthetic aperture radar* sensors.

SIR-A (12–15 November 1981). Orbit: Circular *LEO* at 260 km altitude and inclination 38°. Principal instruments: L-band *SAR*.

SIR-B (5–13 October 1984). Orbit: Circular *LEO* at 225 km altitude and inclination 57°. Principal instruments: L-band *SAR*.

SIR-C/X–SAR (9–20 April 1994; 30 September–11 October 1994; further missions planned). Orbit: Circular *LEO* at 225 km altitude, inclination 57°. Principal instruments: L-band and C-band *SAR*s, *X-SAR*.

SIRS (Satellite Infrared Spectrometer) U.S. nadir-viewing infrared spectrometer for atmospheric temperature profiling, carried on *Nimbus-3* and *-4*. Wavebands:

7 channels between 11 and 15 μm (Nimbus-3); 14 channels between 11 and 36 μm (Nimbus-4). Spatial resolution: 220 km.

SIS   See *Carterra-1*.

Skylab   Series of U.S. space stations, operated by *NASA* between May 1973 and July 1979. Skylabs 2 (May–June 1973), 3 (July–September 1973) and 4 (November 1973–February 1974) were manned. Orbit: Circular *LEO* at 435 km altitude, inclination 108°, period 101 minutes. Principal instruments: *EREP*.

URL: http://www.ksc.nasa.gov/history/skylab/skylab.html

Slant range   The straight-line distance from a *side-looking radar* or *synthetic aperture radar* to a scatterer on the Earth's surface.

SLAR   See *side-looking radar*.

Slick, oil   See *oil spills*.

SLR   See *side-looking radar*.

SM   (Multichannel Spectrometer, also known as Device 174K) Russian thermal infrared radiometer for atmospheric temperature sounding, carried on *Meteor-2* satellites. Wavebands: 9.65, 10.60, 11.10, 13.33, 13.70, 13.25, 14.43, 14.75, 15.02, 18.70 μm. Spatial resolution: 42 km. Swath width: 1000 km.

Small perturbation model   A surface *scattering model* for a random rough surface, relating the scattering coefficient to the frequency, polarisation, incident and scattered angles, surface dielectric constant, surface correlation function and RMS surface height variation. Its conditions for validity are

$k\sigma < 0.3,$

$kL < 3,$

where k is the *wavenumber* of the radiation, σ is the RMS surface height variation, and L is the correlation length of the surface height variation.

The derivation of this model is based on an iterative scheme with the surface height and surface slope as the small parameters. In the current literature, only the first- and second-order results are available. The condition that kL be small does not appear in the derivation, but is needed for the convergence of the perturbation series. The first-order perturbation result takes on the form of *Bragg scattering* and is sometimes called the Bragg scattering model.

The backscattering coefficient is given by this model as

$$\sigma_{pp}^0(\theta) = 8k^4\sigma^2\cos^4\theta|\alpha_{pp}(\theta)|^2 W(2k\sin\theta),$$

where σ_{pp}^0 is the backscattering coefficient for pp-polarisation (i.e. $p = V$ or H), θ is the incidence angle and W is the normalised roughness spectrum, equal to the Bessel transform of the surface height correlation function $\rho(\xi)$. If the surface has a Gaussian autocorrelation function $\rho(\xi) = \exp(-\xi^2/L^2)$,

$$W(2k\sin\theta) = \frac{L^2}{2}\exp(-k^2L^2\sin^2\theta).$$

The term $|\alpha_{HH}|^2$ is just $|R_H(\theta)|^2$, the Fresnel coefficient for intensity reflection of H-polarised radiation at angle θ. α_{VV} is given by the following formula:

$$\alpha_{VV}(\theta) = (\varepsilon - 1)\frac{\sin^2\theta - \varepsilon(1 + \sin^2\theta)}{[\varepsilon\cos\theta + (\varepsilon - \sin^2\theta)^{1/2}]^2},$$

where ε is the (complex) dielectric constant of the surface.
See also *integral equation model, Kirchhoff model*.

SMMR (Scanning Multichannel Microwave Radiometer) U.S. mechanically scanned passive microwave radiometer, carried on *Nimbus-7* and *Seasat* satellites. Frequencies: 6.6, 10.7, 18.0, 21.0, 37.0 GHz. Polarisation: H and V. Spatial resolution: 136×89 km at 6.6 GHz to 28×18 km at 37 GHz. Incidence angle: 50°. Swath width: 780 km (600 km from Seasat). Absolute accuracy: 2 K.

Smoothing A spatial filtering operation performed on an image, usually in order to suppress random noise. Many forms of smoothing operation can be described as *convolution operators*. The simplest is an $N \times N$ average, for example the 3×3 average whose kernel is shown below:

1/9	1/9	1/9
1/9	1/9	1/9
1/9	1/9	1/9

Variants on this type of operator have weights that decrease with increasing distance from the centre of the kernel matrix, although the sum of the weights must still be unity if the process is required not to change the average brightness of the whole image.

These types of smoothing operators are linear, and they have the effect of suppressing high spatial frequencies in the image, for example by smoothing sharp edges and by reducing the contrast between a single bright or dark pixel and its background. (They are therefore sometimes referred to as **low-pass filters**.) If this behaviour is not desired, **non-linear** smoothing operators can be used. For example, the **median filter**, which sets the *digital number* of the pixel with coordinates (i,j) to the median of the N^2 values in an $N \times N$ box centred on (i,j), has reasonable edge-preserving properties, although it does not preserve single bright or dark pixels. An alternative non-linear operator, which does preserve bright or dark pixels, involves comparing the value I' given by a simple $N \times N$ average with the original unsmoothed value I, and using the former only if the difference $|I' - I|$ exceeds some threshold. This threshold value is often calculated using the statistical distribution of the digital numbers within the $N \times N$ box.

SMR (Sub-millimetre radiometer) Swedish passive microwave star-viewing limb sounder planned for inclusion on *ODIN*. Frequencies: 119, 486–504, 541–586 GHz.

The SMR will be used to obtain atmospheric profiles of a wide range of molecular species.

SMS See *GOES*.

Snell's laws Laws governing the directions of the reflected and transmitted rays when plane parallel radiation is incident at a planar interface between two media. If the radiation is incident from within medium 1 and makes an angle θ_1 with the normal to the interface, the reflected ray is reflected specularly at angle θ_1 and the transmitted ray is refracted at angle θ_2, where

$$n_1 \sin \theta_1 = n_2 \sin \theta_2,$$

n_1 and n_2 are the refractive indices of the two media, and the three rays are coplanar, as shown in the figure.

See also *Fresnel coefficients*.

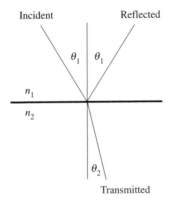

Snow Solid *precipitation* (ice crystals) falling through the air or deposited on the ground. Snow represents approximately 5% of all precipitation reaching the Earth's surface. Its climatological significance is derived mainly from its high *albedo* and its low thermal conductivity. Snowfalls which do not immediately melt may build a snow pack which can have economic importance as a water resource.

The terms **snow cover** or **snow extent** can be used interchangeably to denote the amount of snow on the ground at the time of observation. The ground may be completely or partially covered by snow. The area of land covered by a recent or old snowfall is important both hydrologically and climatologically. In terms of hydrology, the snow-covered area is directly related to snowmelt runoff, and hence stream flow, of many river basins. Since the atmosphere is sensitive and responsive to changes in its lower boundary, snow cover can affect climate at local, regional and global scales. Snow cover can also be an important restraint on plant and animal (domesticated and wild) life, limiting the regions where flora and fauna can establish themselves, and insufficient snow cover in agricultural areas can cause the death, through excessive cold, of autumn-planted crops.

Snow cover has been mapped, monitored and measured from space for the past three decades. Because of the large contrast between snow-covered and snow-free terrain at visible wavelengths, those sensors operating in the reflective portion of the solar spectrum, such as the *NOAA* series of satellites, have been extensively used to map snow. Increasing use is now also being made of *synthetic aperture radar* observations, which can provide good discrimination between wet and dry snow surfaces.

See also *glaciers*.

Snow, electromagnetic properties

Visible and near infrared region
The following table shows the approximate value of the diffuse *albedo* of <u>dry snow</u>, in per cent, from 0.4 μm to 1.6 μm.

	Wavelength (μm)												
Grain size (μm)	0.4	0.5	0.6	0.7	0.8	0.9	1.0	1.1	1.2	1.3	1.4	1.5	1.6
50	98	98	98	97	95	90	82	80	68	64	50	10	13
200	98	98	98	95	90	80	65	65	54	42	33	2	3
1000	97	95	93	88	79	60	40	41	20	16	10	0	0

Thermal infrared region
The *emissivity* of dry snow is between 0.8 and 1.0 at wavelengths 8–14 μm.

Microwave region
The *dielectric constant* of <u>dry snow</u> is a function only of its density. The real part is given approximately by the formulae

$$\varepsilon' = 1 + 0.0019\rho \quad [\rho < 500],$$

$$\varepsilon' = 0.51 + 0.0029\rho \quad [\rho \geq 500],$$

where ρ is the density in $kg\,m^{-3}$. The imaginary part also depends on the frequency and the temperature, and is given reasonably accurately for frequencies up to about 15 GHz by the formula

$$\varepsilon'' = 0.0016(5.2 \times 10^{-4}\rho + 6.2 \times 10^{-7}\rho^2)(f^{-1} + 0.389f^{1/2})\exp(0.036T),$$

where f is the frequency in GHz and T is the temperature in °C. The attenuation coefficient is normally dominated by scattering (except at low frequencies), and depends strongly on the grain size as well as the frequency. For dry snow at frequencies between about 18 and 60 GHz it is given approximately by

$$0.0018f^{2.8}d^2 \text{ dB/m},$$

where d is the grain size in mm.

For <u>wet snow</u>, the dielectric constant can be obtained from the following formulae, due to Hallikainen and Winebrenner and based on modified Debye equations (other formulations for the dielectric constant of wet snow are possible, notably two-phase mixing models):

$$\varepsilon' = A + \frac{Bw^{1.31}}{1 + (f/f_0)^2},$$

$$\varepsilon'' = \frac{C(f/f_0)w^{1.31}}{1 + (f/f_0)^2},$$

where f_0 is the appropriate relaxation frequency (9.07 GHz), A, B and C are dependent on frequency and density, and w is the volume fraction (in per cent) of liquid water. For frequencies between 1 and 15 GHz, A, B and C are given empirically by

$$A = 1 + 0.00183\frac{\rho - 10w}{1 - \dfrac{w}{100}} + 0.02w^{1.015},$$

$$B = C = 0.073.$$

For frequencies between 15 and 40 GHz,

$$A = 1 + 0.00183\frac{\rho - 10w}{1 - \dfrac{w}{100}} + 0.02Pw^{1.015} + Q,$$

$$B = 0.073P,$$

$$C = 0.073R,$$

where

$$P = 0.78 + 0.03f - 5.8 \times 10^{-4}f^2,$$

$$Q = 0.31 - 0.05f + 8.7 \times 10^{-4}f^2,$$

$$R = 0.97 - 3.9 \times 10^{-3}f + 3.9 \times 10^{-4}f^2.$$

Snow line See *glaciers*.

Snow volume The volume of a snow pack in a given area, for example a watershed, is most usefully expressed as the equivalent volume of water that would be released if all the snow were melted. This is determined by estimating the area (see *snow cover*), the average depth and the average density of the snow. The density of snow can vary between 100 kg m^{-3} for newly fallen snow, to about 800 kg m^{-3} for old, compacted snow, though a more typical value for a mature snow pack in spring is 500 kg m^{-3}. In practice, the snow density is usually assessed by dividing the *snow water equivalent* by the depth. Snow depth values can be estimated remotely, using multifrequency *passive microwave* data, but are still more commonly estimated by *in situ* measurement. Passive microwave estimates of snow depth rely on the large dielectric contrast between snow and dry soil, and the attenuation of the signal emitted by the ground as it travels through the snow pack. For example, the following empirical formula has been used to estimate dry snow depths up to about 100 cm using horizontally polarised passive microwave $SMMR$ brightness temperatures at 18 and 37 GHz, assuming a snow density of $300\,\mathrm{kg\,m^{-3}}$ and a grain size of 0.35 mm:

$$d/\mathrm{cm} = 1.59(T_{18H} - T_{37H}).$$

Snow water equivalents can also be estimated from visible-wavelength or near infrared *albedo*, if the snow lies over a low-albedo surface, again through the effect of attenuation of the signal in the snow pack. The following table shows the visible-wavelength reflectance of snow overlying a surface of zero albedo, as a function of snow water equivalent s, grain size d and solar zenith angle θ.

s (mm)	d (μm)	$\theta = 30°$ 50	200	1000	$\theta = 60°$ 50	200	1000
0.5		0.52	0.18	0.04	0.63	0.31	0.09
1.0		0.70	0.33	0.07	0.77	0.47	0.17
2.0		0.82	0.50	0.14	0.88	0.64	0.30
5.0		0.92	0.73	0.32	0.94	0.79	0.45
10		0.96	0.85	0.50	0.97	0.88	0.62
20		0.97	0.91	0.67	0.98	0.94	0.76
50		0.97	0.94	0.82	0.98	0.96	0.86
100		0.97	0.94	0.86	0.98	0.96	0.90
200		0.97	0.94	0.86	0.98	0.96	0.91

Snow water equivalent Snow water equivalent is the depth of water that would result if the snow were to melt (see *snow volume*). It is often measured *in situ*, but can be estimated from multifrequency passive microwave observations.

SNSB (Swedish National Space Board) See *CEOS*.

Sobel operator See *edge detection*.

Soil, electromagnetic properties
Visible and near infrared region
The reflectance properties of soils are strongly influenced by soil type (mineral and organic matter contents) and by *soil moisture*. In the optical band, reflectance normally increases with increasing wavelength, but can be sufficiently characteristic of the soil type that the soil colour can be used for identification. Spectrometry in the optical and near infrared bands can reveal the characteristic absorption features of particular ions and of water.

Thermal infrared region
The *emissivity* of soils in the 8–14 μm band is typically 0.88 for sandy soils to 0.98 for clays and loams. The emissivity of sandy soils increases with increasing

water content, reaching a value of about 0.95 at a volumetric moisture content of 10%.

Typical emissivities in the 3–5 and 8–14 μm regions:

	3–5 μm	8–14 μm
clay loam	0.88	0.97
sand, wet or dry		0.90–0.93
silt loam	0.74–0.85	0.93–0.98
soil, dry		0.92
soil, wet		0.96

Microwave region

The *dielectric constant* of soil is determined by the soil type and *soil moisture*, and is dependent upon the frequency. Dry soils usually have $\varepsilon' \approx 3$ and $\varepsilon'' \approx 0$ at frequencies below about 20 GHz. The table below shows representative values for sandy, clay and moist loam soils as functions of the volumetric water content w.

f (GHz)	w (%)	sand ε'	ε''	clay ε'	ε''	loam ε'	ε''
1.0						10	1
1.4	0	3	0	3	0		
1.4	20	11	1	8	2		
1.4	40	25	2	20	5		
1.4	60	50	3	40	7		
5	0	3	0	3	0		
5	20	9	1	6	1		
5	40	17	2	17	4		
5	60	18	3	25	7		
10						10	5

Soil adjusted vegetation index See *vegetation index*.

Soil and moisture marks in archaeology The soil's mechanical structure and its effects on surface colour and water retention are significant. Materials transported by man from lower layers may be visible from the air on a bare surface, giving rise to the soil site. Spectral reflectance is a property of the material, and a function of moisture, organic matter, iron oxide content and texture. Since relative lightness is more important than position in the spectrum, colour changes may be enhanced in black and white aerial photographs. Colour is also affected by grain size distribution, but surface roughness does not affect appearance as much, although the direction of ploughing is important. Foreign materials such as brick or stone may often remain on the surface improving visibility. Structures visible as colour differences can be seen for only short periods until rain, wind and agricultural treatment obliterate them. Sites

on shallow soils or slopes are favoured, since the possibility of material being brought to the surface is higher. Early spring, just after ploughing, planting and rolling is optimal for recording sites in diffuse light. Outlines are usually rather blurred unless large mixed material is brought to the surface.

See also *archaeological site detection, soil moisture*.

Soil moisture Soil moisture estimates are needed for agricultural applications, hydrology and climatology, and are used to estimate the rates of evaporation, surface runoff, and percolation to the water table. Consequently, soil moisture measurements find applications in the prediction of crop yields (and hence in famine management and global food security), irrigation management, and monitoring areas liable to erosion and *desertification*. Although most soil moisture measurements are still acquired by ground survey, various remote sensing techniques can give reasonable estimates. Optical and near infrared reflectance is strongly correlated with soil moisture (increasing water content lowers the reflectance), but only for the top few millimetres of the soil. Indirect estimates of soil moisture can be made if the type and condition of *vegetation* is known (e.g. from the analysis of optical/near infrared imagery) and combined with the land surface temperature, determined from thermal infrared measurements.

Daytime and night-time thermal infrared measurements can be combined to determine the amplitude of the diurnal temperature fluctuations and hence deduce the *thermal inertia*, which is strongly dependent on soil moisture content. This technique can not be applied when the soil surface is vegetated or obscured by cloud, and increasing use is therefore now being made of microwave techniques. The microwave *emissivity* of a soil surface is correlated with its moisture content, so passive microwave measurements can be used to infer soil moisture. Low frequencies are preferable, since they provide better penetration of vegetation cover and also sample a greater depth of the soil, but the use of low frequencies reduces the *spatial resolution*. Synthetic aperture radar measurements overcome this limitation. The *backscattering coefficient* of the soil surface is dependent upon its dielectric properties (which are largely determined by moisture) and by the surface and volume structure of the soil. However, the presence of a vegetation cover causes further scattering which masks the signal from the soil and hence reduces the sensitivity of the method to variations in soil moisture content.

Solar Backscatter UV Radiometer See *SBUV*.

Solar constant See *solar radiation*.

Solar illumination direction The direction from which sunlight illuminates the Earth's surface is significant for remote sensing systems which detect reflected sunlight. This direction is normally specified by the Sun's elevation angle *a* and azimuth angle *A*. The elevation angle is the angle that the line of sight to the Sun makes with a horizontal plane; the azimuth angle is the bearing of the line of sight (north = 0°, east = 90° etc.). These angles can be calculated from the formulae

$$\sin a = \sin \delta \sin \phi + \cos \delta \cos \phi \cos H,$$

$$\cos A = \frac{\sin \delta - \sin \phi \sin a}{\cos \phi \cos a},$$

where δ is the Sun's **declination**, ϕ is the latitude (north positive) of the illuminated point, and H is the Sun's hour angle. Note that the formula for A contains an ambiguity in taking the inverse cosine. This ambiguity is resolved by noting that A and H must have the same sign after they have both been reduced to the range from $-180°$ to $+180°$ by repeated addition or subtraction of multiples of $360°$. The hour angle, in degrees, is given by the formula

$$H = 15T - 180 + \lambda + E/4,$$

where T is the Greenwich Mean Time in hours, λ is the longitude in degrees (east positive), and E is the **equation of time** in minutes.

The values of δ and E depend only on the date, and repeat very closely from year to year. The figure shows their variation throughout the year.

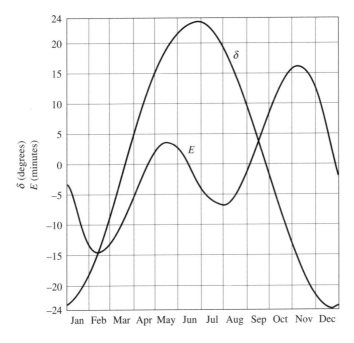

Solar radiation Electromagnetic radiation from the Sun can be approximately characterised as *black-body radiation* at a temperature of about 5800 K, with very many absorption lines (Fraunhofer lines) due to atoms and ions in the Sun's outer layers. The **solar constant** is defined as the *irradiance* at a distance equal to the Earth's mean distance from the Sun. It has a value of approximately 1.37 kW m^{-2}.

Sounder See *atmospheric sounding*.

Spacelab Spacelab missions were short missions of the *Space Shuttle* with various objectives including remote sensing.

Spacelab-1 (28 November to 8 December 1983). Orbit: nominally circular *LEO* at 240 to 260 km altitude, inclination 57°. Principal instruments: *MC*. Spacelab-1 also carried a microwave radiometer/scatterometer, an infrared spectrometer for profiling atmospheric molecular species, and a solar irradiance spectrometer.

Spacelab-3 (29 April to 6 May 1985). Orbit: nominally circular *LEO* at 357 km altitude, inclination 57°. Principal instruments: *ATMOS*.

Space segment See *ground segment*.

Space Shuttle U.S. reusable, manned, launch vehicles that can carry a large payload into *LEO*. The first Space Shuttle, also called STS (Space Transportation System), was launched in December 1981. Space Shuttle missions are used for many purposes including remote sensing.

URL: http://shuttle.nasa.gov/

Space Transportation System See *Space Shuttle*.

Spatial dependence matrix See *grey level co-occurrence matrix*.

Spatial filtering See *convolution operator*.

Spatial frequency See *Fourier transform*.

Spatial resolution A measure of the smallest distance between two objects that can be distinguished by a sensor. Various definitions of spatial resolution are possible, depending on what is meant by the words 'object' and 'distinguish'. For an *electro-optical sensor*, the simplest definition is the instantaneous field of view (IFOV), which is the area of the imaged surface that is projected, through the sensor's optical system, onto a single detector element. This will normally correspond to the area of a *pixel* in the image, although the pixel may be smaller if the image is over-sampled. The IFOV can be specified as an angular width, or as a spatial extent in which case it is dependent on the distance from the target to the sensor as well as on the properties of the sensor itself.

The spatial resolution can also be defined by the point-target response due to diffraction and aberration effects in the optics of a photographic or electro-optical system or at the antenna of a microwave system. For a perfect (aberration-free) circular aperture of diameter D, the diffraction-limited angular resolution for radiation of wavelength λ incident normally on the aperture is approximately $1.2\lambda/D$ radians, provided that $D \gg \lambda$. If this resolution is significantly coarser than the IFOV, radiation from a point target will be detected by several pixels instead of just one, thus degrading the useful spatial resolution. This is the dominant factor in determining the spatial resolution of *passive microwave* systems. For *imaging radar* systems, somewhat different considerations apply. The across-track (range) resolution is determined by the duration of the radar pulse. The along-track (azimuth) resolution of an *SLR* system is determined by diffraction effects, but for a *synthetic aperture radar* system it is set by the maximum length of the synthetic aperture over which coherent data can be collected, or by the manner in which the data are processed.

A third definition of spatial resolution, normally used for photographic systems, relates to the ability to detect periodic targets. The resolution is specified as the maximum number of line-pairs per unit length (usually per millimetre) detectable when the target consists of alternating black and white strips of equal width. The resolution is specified with respect to the film plane, so a system having a focal length of f mm and a resolution of l line-pairs per millimetre can resolve a periodic target with an angular period of $1/lf$ radians. This is roughly equivalent to a point-target resolution of $1/2lf$ radians, although a more detailed analysis involves the use of the *modulation transfer function.*

The effective spatial resolution of an image depends on the use to which it is put, and will be affected not only by the parameters of the observing system but also by the properties of the target itself. The least demanding task, in terms of the requirement for spatial resolution, is the simple detection of the presence of some object. If the target has a high radiometric contrast with its background, its presence can be detected even if its size is less than the IFOV, provided that the sensor has sufficient *radiometric resolution*. The most demanding task is the spectral or spatial analysis of the properties of the target, in which case it must contain many pixels.

Somewhat different considerations apply in the case of non-imaging systems such as *radar altimeters, laser profilers* and *atmospheric sounders*, where both the horizontal and vertical (height) resolution must be considered.

Special Sensor C See *SSC.*

Special Sensor D See *SSD.*

Special Sensor H See *SSH.*

Special Sensor Microwave Imager See *SSM/I.*

Special Sensor Microwave Temperature Sounder See *SSM/T.*

Specific impulse See *orbital manoeuvres.*

Speckle Multiplicative noise introduced by any *coherent* imaging system, for example *synthetic aperture radar*, imaging a non-specular surface. The signal detected from a uniform homogeneous area can be assumed to be made up of many components having the same amplitude and random phase. When these components are combined coherently, i.e. having regard to both amplitude and phase, the resultant amplitude A has a probability distribution given by the Rayleigh distribution:

$$p(A) = \frac{\pi A}{2\langle A\rangle^2} \exp\left(-\frac{\pi A^2}{4\langle A\rangle^2}\right),$$

where A is positive or zero, and $\langle A\rangle$ is the expectation value of A. The conventional intensity I (strictly the square of the amplitude) has an expectation value of

$$\frac{4\langle A\rangle^2}{\pi}$$

and a probability distribution given by a negative exponential distribution:

$$p(I) = \frac{\pi}{4\langle A\rangle^2} \exp\left(-\frac{\pi I}{4\langle A\rangle^2}\right).$$

If the signal is expressed in decibels as $S = 10\log_{10}(I) = 20\log_{10}(A)$, the expectation value $\langle S\rangle$ is $20\log_{10}\langle A\rangle - 1.5\,\text{dB}$ and the standard deviation of S is about $5.6\,\text{dB}$.

The effects of speckle in synthetic aperture radar imagery are often reduced by *multi-look processing* and/or some other form of spatial averaging.

Spectral radiance Spectral radiance L_λ or L_ν is defined such that the *radiance L* between wavelengths λ_1 and λ_2 is given by

$$L = \int_{\lambda_1}^{\lambda_2} L_\lambda \, d\lambda$$

and the radiance between frequencies ν_1 and ν_2 is given by

$$L = \int_{\nu_1}^{\nu_2} L_\nu \, d\nu.$$

L_λ and L_ν are related by the formula

$$\frac{L_\lambda}{L_\nu} = \frac{c}{\lambda^2} = \frac{\nu^2}{c},$$

where c is the speed of light.

Spectral resolution The ability of a sensor to discriminate between different wavelengths (or frequencies) in the detected signal.

In the visible and near infrared regions of the electromagnetic spectrum, remote sensing instruments can have spectral resolutions ranging from more than $1\,\mu\text{m}$ to less than $1\,\text{nm}$. The broadest-band sensors are the *imaging radiometers*, which generally have high *spatial resolution* and/or *radiometric resolution*.

Most *multispectral imagers* operate at a few wavebands, defined by filters to have bandwidths of typically 50 to 100 nm. These bands are not necessarily contiguous. Higher spectral resolutions, typically 20 nm, are needed for *ocean colour* instruments, and these are also provided by filters. Instruments that provide yet higher spectral resolution, in contiguous bands, are referred to as spectrometers, imaging or otherwise. Resolutions of the order of 0.1 nm in the optical and near infrared, sufficient to resolve many spectral lines, are generally obtained using diffraction gratings to disperse the spectrum onto an array of detectors.

Similar considerations apply to the thermal infrared region. Broad-band measurements, with filter-defined bandwidths of the order of 1 μm, are generally used for surface temperature measurements. Instruments for *atmospheric sounding* need much higher spectral resolution in order to resolve spectral lines. This is often achieved by Fourier transform spectrometry, in which a Michelson interferometer is used to resolve the different wavelengths present in a broad-band signal. Since the spectral resolution of such an instrument is constant when measured in terms of *wavenumber* (or equivalently frequency), it is proportional to the square of the wavelength. Typical resolutions range from 0.1 nm at 3 μm to 2 nm at 15 μm.

An alternative approach to obtaining high spectral resolution in the thermal infrared region, used in some atmospheric *limb sounders*, is gas cell correlation radiometry. The instrument carries a cell containing the gas to be detected. The pressure in the cell is varied over the range of interest, and the intensity of radiation detected from the Sun when looking through the cell but not through the atmosphere is compared with the signal observed by looking through the atmosphere but not through the cell.

At passive microwave frequencies, both sounders and radiometers use electronic filtering to achieve spectral resolutions of the order of 0.1–1 GHz.

Obtaining increased spectral resolution requires either that the detected signal is dispersed into many spectral components, which has the effect of decreasing the signal and hence degrading the *radiometric resolution*, or that a longer time is devoted to the measurement, which has the effect of decreasing the *spatial resolution*.

Spectroradiometer for Ocean Monitoring See *SROM*.

Specular reflection A perfectly smooth surface (see *Rayleigh criterion*) scatters radiation specularly, i.e. like a mirror. The *bidirectional reflectance distribution function* of such a surface is given by

$$R(\theta_0, \phi_0, \theta_1, \phi_1) = \frac{2|r_p(\theta_0)|^2 \delta(\theta_1 - \theta_0)\delta([\pi + \phi_1 - \phi_0])}{\sin 2\theta_0},$$

where $r_p(\theta_0)$ is the *Fresnel coefficient* for radiation of polarisation state p incident at angle θ_0, $\delta(x)$ is the Dirac delta-function, and $[x]$ denotes addition or subtraction of an integer multiple of 2π from x to bring it into the range $-\pi \leq x \leq \pi$.

Spektr See *Mir-1*.

Spin-scan imager The type of optical and infrared remote sensing system most commonly used on satellites in *geostationary* orbit, to acquire imagery from the whole of that part of the Earth's surface visible from the satellite's position. Side-to-side scanning (scan lines parallel to the equator) is achieved by arranging for the whole satellite to spin, typically at 100 r.p.m., about its vertical axis. Vertical (north–south) scanning is achieved using an oscillating mirror. The time required for such an instrument to acquire a complete image depends on the spin rate and the number of scan lines, and is typically about 30 minutes.

SPOT (Système Probatoire d'Observation de la Terre, or Satellite Pour l'Observation de la Terre) European (France, Sweden and Belgium) satellite system operated by *CNES*. SPOT-1 was launched in February 1986 and decommissioned in December 1990 (reactivated in January 1997 but without on-board storage), SPOT-2 was launched in January 1990 (on-board storage failed September 1993), SPOT-3 in September 1993 (mission ended November 1996) and SPOT-4 in March 1998. SPOT-5 is scheduled for launch in 2002. Objectives: Land surface mapping and monitoring. Orbit: Circular *Sun-synchronous LEO* at 822 km altitude. Period 101.5 minutes; inclination 98.7°; equator crossing time 10:30 (descending node). *Exactly repeating orbit* (369 orbits in 26 days). Principal instruments: *HRG* (SPOT-5), *HRV* (SPOT-1 to -3), *HRVIR* (SPOT-4), *POAM-2* (SPOT-3), *POAM-3* (SPOT-4), *Vegetation* (SPOT-4). SPOT-2 onward also carry the *DORIS* satellite location system.

URL: http://www.spotimage.fr/

SR (Scanning Radiometer) U.S. ultraviolet/optical/infrared mechanically scanned imaging radiometer, carried on *TIROS*, *ITOS* and *NOAA* satellites. Wavebands: 0.2–0.6, 0.5–0.75, 6.0–6.5, 8.0–12.0, 8.0–13.0 μm. Spatial resolution: 3.2 km (8 km for thermal infrared). Swath width: 2500 km.

SROM (Spectroradiometer for Ocean Monitoring) Russian optical/infrared mechanically scanned imaging radiometer, planned for inclusion on *Almaz-1B*. Wavebands: 405–422, 433–453, 480–500, 510–530, 555–575, 655–675, 745–785, 843–884 nm; 3.6–3.9, 10.5–11.5, 11.5–12.5 μm. Spatial resolution: 600 m at nadir. Swath width: 2200 km.

Temperature sensitivity (thermal infrared bands): 0.1 K. The optical/near infrared bands are primarily intended for ocean colour measurements.

SSALT (Single-frequency Solid State Altimeter) French nadir-viewing radar altimeter, carried by *Topex-Poseidon*. Frequency: K_u band (13.5 GHz). Height precision: 2.5 cm.

URL: http://www-aviso.cls.cnes.fr/English/TOPEX_POSEIDON/More_On_Payload.html

SSC (Special Sensor C) U.S. single-band infrared imaging radiometer, carried on F4 satellite of *DMSP* Block 5D-1 series. Waveband: $1.51-1.63\,\mu m$. Spatial resolution: 12 km. Swath width: 650 km.
Experimental sensor to investigate the discrimination between snow and cloud.

SSD (Special Sensor D) U.S. ultraviolet limb sounder, for vertical profiling of atmospheric N_2, O_2 and O_3, carried on *DMSP Block 5D-1* satellites. The SSD measured backscattered solar radiation.

SSE (Supplementary Sensor E) Infrared radiometer for vertical profiling of atmospheric temperature and water vapour, carried on *DMSP* block 5B/C satellites. Eight wavebands in the range $11\,\mu m$ to $15\,\mu m$. Horizontal resolution 37 km, swath width 185 km. Also known as VTPR (Vertical Temperature Profile Radiometer).

SSH (Special Sensor H) U.S. scanning infrared spectrometer, for measuring temperature, water vapour and ozone concentrations in the atmosphere, carried on *DMSP* Block 5D-1. Wavebands: 16 channels between 9.6 and $30\,\mu m$ (CO_2, H_2O and O_3 absorption lines). Spatial resolution: 39 km. Swath width: 2200 km.
The SSH is also known as VTPR (Vertical Temperature Profiling Radiometer). DMSP Block 5D-2 satellites carry an improved instrument, SSH-2, with narrower spectral bands, an increased spectral range ($3.7-30\,\mu m$), and a spatial resolution of 60 km.

SSM/I (Special Sensor Microwave Imager) U.S. mechanically scanned passive microwave radiometer, carried on *DMSP* Block 5D-2 satellites. Frequencies: 19.35, 22.24, 37.0, 85.5 GHz. Polarisation: V (all frequencies), H (all frequencies but 22.2 GHz). H channel on 85.5 GHz band of instrument on satellite F8 failed in February 1990. Spatial resolution: $70\,km \times 45\,km$ at 19.4 GHz to $16 \times 14\,km$ at 85.5 GHz. Swath width: 1400 km. Sensitivity: 0.8 K. Absolute accuracy: 1.5 K.
DMSP block 5D-3 satellites will carry SSM/IS instruments which include two atmospheric sounding channels.

SSM/T (Special Sensor Microwave Temperature Sounder) U.S. scanning passive microwave radiometer for atmospheric temperature profiling, carried on *DMSP* Block 5D-1 and Block 5D-2 satellites. Frequencies: 50.5, 53.2, 54.35, 54.9, 58.4, 58.83, 59.4 GHz. Spatial resolution: 175 km at nadir, Swath width: 1500 km. Temperature resolution: 0.3-0.5 K. Height resolution: 15 levels between 0 and 30 km.
Satellite F-11 carries the SSM/T-2 instrument with channels at 91.5, 150 and 183 GHz for water vapour sounding. It has a spatial resolution of 50 to 120 km depending on frequency.

SSR-1 (Satelite de Sensoriamento Remoto) Brazilian satellite, operated by *INPE*, scheduled for launch in 1999 with a nominal lifetime of 4 years. Objectives: Land surface monitoring, particularly vegetation and particularly in Brazil. Orbit: Circular *LEO* at 900 km altitude. Period 97.6 minutes; inclination 97.9°; equator crossing time 09:30. Principal instruments: *WFI*.

URL: http://www.met.inpe.br/wwwmecb/ssr.htm

SSU (Stratospheric Sounding Unit) U.K. nadir-viewing infrared radiometer (pressure-modulated gas cell) for stratospheric temperature profiling, carried on *TIROS-N* and *NOAA-6* onwards (as part of the *TOVS* package). Wavebands: 14.926, 14.934 and 14.940 μm. Spatial resolution: 150 km.

Starlette French satellite for solid-earth research. Starlette consists of an icosahedral array (12 cm diameter) of laser retroreflectors which can be ranged from laser tracking ground stations to give Earth gravity field data. Semimajor axis: 7330 km. Eccentricity: 0.02. Inclination: 50°. See *Stella*.

Stationary phase model See *Kirchhoff model*.

Station mask The region of space from which signals can be received by a *receiving station*. The station mask is usually represented as the region within which the *sub-satellite point* of a satellite at height h must lie in order for signals from the satellite to be detectable at the receiving station. If the receiving station antenna is fully steerable and has sufficient sensitivity, the limiting factor is the physical obstruction provided by the Earth's surface. For a receiving station situated at ground level with an unobstructed horizon, the station mask is a small circle of radius r, centred on the receiving station, where

$$r = R \arccos \frac{R \cos \theta}{R + h} - R\theta,$$

R is the Earth's radius and θ is the minimum acceptable elevation angle of the line of sight (typically 0.1 radians). In this expression, angles are expressed in radians.

In the case of a spaceborne remote sensing instrument that views towards nadir and transmits image data directly to a receiving station, the station mask also corresponds to the area on the Earth's surface from which data can be obtained by that receiving station.

Stefan's law The *radiant exitance* of a *black body* at absolute temperature T is σT^4, where σ is the Stefan–Boltzmann constant.

Stella French satellite for solid-earth research, launched September 1993 with a design lifetime of 10 000 years. Stella consists of an icosahedral array (12 cm diameter) of laser retroreflectors which can be ranged from laser tracking ground stations to give Earth gravity field data. Semimajor axis: 7160 km. Eccentricity: 0. Inclination: 98°. See *Starlette*.

Step-stare imager *Imaging system* using a two-dimensional *CCD* to acquire a two-dimensional image. The instrument 'stares' at the target scene (maintains a constant viewing geometry) to collect radiation from it, then 'steps' to the next field of view. Each detector element of a step-stare imager views its target *rezel* for much longer than is possible for a *pushbroom* or *whiskbroom* sensor, giving an improved signal-to-noise ratio. The disadvantage is the difficulty of calibrating the very large number (of the order of a million) of detector elements.

Stereophotography An aspect of *photogrammetry* relating to the determination of surface topography from overlapping vertical aerial photographs. A single vertical aerial photograph will image a target point located at (x, y, h) at

$$\left(\frac{fx}{H-h}, \frac{fy}{H-h} \right),$$

where x and y are the horizontal Cartesian coordinates of the target point, measured in two orthogonal directions from the nadir point (assumed coincident with the *principal point*), h is the height of the target point above a horizontal reference plane, f is the focal length of the camera, and H is the camera's height above the reference plane. (See *relief displacement.*) The three coordinates x, y and h of the target point are thus represented as only two coordinates in the photograph, and therefore cannot be determined uniquely unless more information is available. Stereophotography provides the necessary information by using a second photograph taken from a different location. In essence, it uses the parallax between observations acquired from the two locations to determine the range to the target point.

In its simplest and commonest form, both photographs are vertical and are acquired from the same height H. The separation B between the two camera positions is termed the **baseline**. The two camera positions can be written in Cartesian coordinates as $(0, 0, H)$ and $(B, 0, H)$, such that the point (x, y, h) will be imaged at (X_1, Y_1) with respect to the centre of the first photograph, where

$$X_1 = \frac{fx}{H-h}$$

and

$$Y_1 = \frac{fy}{H-h}$$

as before. Assuming that the point is also visible in the second photograph, its image coordinates (X_2, Y_2) will be

$$X_2 = \frac{f(x-B)}{H-h}$$

and

$$Y_2 = \frac{fy}{H-h}.$$

The target coordinates x, y and h can thus be determined uniquely as

$$x = \frac{X_1 B}{X_1 - X_2},$$

$$y = \frac{Y_1 B}{X_1 - X_2},$$

$$h = H - \frac{fB}{X_1 - X_2}.$$

The quantitative application of these formulae is normally accomplished manually, using a suitable optical instrument. The instrument is initially set to the viewing parameters of the two photographs. The operator then makes adjustments to bring the two images of a target point into coincidence, when the instrument will indicate the corresponding height. Increasing use is now being made of automated stereo matching, in which photographs are first scanned into a computer and then matched by spatial correlation.

The accuracy with which heights can be determined by stereophotography is increased by increasing the baseline B. However, this also has the effect of decreasing the area of the target that is common to both photographs, and in practice stereo pairs are normally acquired with an overlap of approximately two thirds of the photograph's width. For a typical mapping system this gives a vertical height resolution of roughly $H/1000$.

STIKSCAT See *SeaWinds*.

Stokes vector A means of specifying the state of *polarisation* of electromagnetic radiation. If the radiation is propagating in the z-direction, the electric field vector rotates parallel to the xy-plane. Its components in the x- and y-directions can be written as

$$E_x = E_{0x} \cos(\langle k \rangle z - \langle \omega \rangle t + \delta_x),$$

$$E_y = E_{0y} \cos(\langle k \rangle z - \langle \omega \rangle t + \delta_y),$$

where $\langle k \rangle$ is the mean wavenumber, $\langle \omega \rangle$ the mean angular frequency, and the amplitudes E_{0x}, E_{0y} and phase angles δ_x, δ_y can vary with time. The Stokes vector is written as a column vector:

$$\begin{bmatrix} S_0 \\ S_1 \\ S_2 \\ S_3 \end{bmatrix} = \begin{bmatrix} \langle E_{0x} \rangle^2 + \langle E_{0y} \rangle^2 \\ \langle E_{0x} \rangle^2 - \langle E_{0y} \rangle^2 \\ \langle 2E_{0x}E_{0y} \cos(\delta_y - \delta_x) \rangle \\ \langle 2E_{0x}E_{0y} \sin(\delta_y - \delta_x) \rangle \end{bmatrix}.$$

The angle brackets $\langle \ \rangle$ denote time-averages. The degree of polarisation (fraction of the total power contained in polarised components) is given by

$$\frac{(S_1^2 + S_2^2 + S_3^2)^{1/2}}{S_0}.$$

The Stokes vectors of incoherent components are additive, and can thus be used to express the polarisation state of incoherent radiation in terms of combinations of simpler components. Examples of Stokes vectors are given below, normalised to $S_0 = 1$:

$$
\begin{array}{ccccccc}
\text{random} & \text{linear-}x & \text{linear-}y & \text{linear }+45° & \text{linear }-45° & \text{RH-circular} & \text{LH-circular} \\
\begin{bmatrix} 1 \\ 0 \\ 0 \\ 0 \end{bmatrix} &
\begin{bmatrix} 1 \\ 1 \\ 0 \\ 0 \end{bmatrix} &
\begin{bmatrix} 1 \\ -1 \\ 0 \\ 0 \end{bmatrix} &
\begin{bmatrix} 1 \\ 0 \\ 1 \\ 0 \end{bmatrix} &
\begin{bmatrix} 1 \\ 0 \\ -1 \\ 0 \end{bmatrix} &
\begin{bmatrix} 1 \\ 0 \\ 0 \\ 1 \end{bmatrix} &
\begin{bmatrix} 1 \\ 0 \\ 0 \\ -1 \end{bmatrix}.
\end{array}
$$

See also *Poincaré sphere*.

STR (Scanning TV Radiometer) Russian optical/infrared spin-scan imaging radiometer, carried on *GOMS* satellite. <u>Wavebands</u>: 0.4–0.7, 6.7–7.0, 10.5–12.5 μm. <u>Spatial resolution</u>: 1.5 km (optical), 6.5 km (infrared), at nadir. <u>Field of view</u>: Full Earth disc as seen from *geostationary* orbit. <u>Scan time</u>: 30 minutes (full disc).

Stratosphere One of the main divisions of the *atmosphere*, lying between the tropopause (typically 8 km above sea level at the poles and 18 km in the tropics) and the stratopause (typically 50 km above sea level). The temperature in the lower half of the stratosphere is constant at about 217 K (-56 °C), rising to about 273 K (0 °C) at the stratopause as a result of absorption of solar ultraviolet radiation by ozone (itself formed as a result of ultraviolet excitation of oxygen).

Stratosphere and Mesosphere Sounder See *SAMS*.

Stratospheric Aerosol and Gas Experiment See *SAGE*.

Stratospheric Aerosol Measurement See *SAM II*.

Stratospheric Sounding Unit See *SSU*.

Striping See *banding*.

STS See *Space Shuttle*.

Subcycle See *exactly repeating orbit*.

Sub-millimetre radiometer See *SMR*.

Sub-satellite point The point vertically below a satellite in orbit about the Earth (i.e. the point where the line joining the satellite to the Earth's centre intersects the surface). Also called the nadir point.

Sun glint *Specular reflection* of sunlight from the Earth's surface (usually from the sea surface). Sun glint is generally undesirable but can contain useful information on the surface roughness, including wind-induced anisotropy. Also called Sun glitter.

Sun-synchronous orbit An *orbit* in which the rate of *precession* is equal in magnitude and sense to the angular velocity of the Earth's centre about the Sun. The effect of this is that plane of the orbit always makes the same angle with the line joining the Earth's centre to the Sun. The *sub-satellite point* of a satellite in a circular Sun-synchronous orbit will cross a given latitude in a given sense (i.e. northbound or southbound) at the same *mean local solar time* regardless of longitude or date. For this reason, such orbits allow the solar illumination characteristics to be standardised to a large extent, and they are widely used for remote-sensing satellites. (See *solar illumination direction.*)

For a circular Sun-synchronous orbit, which is necessarily *retrograde*, the *inclination i* and *nodal period* P_n are both determined by the orbital radius a. The table below illustrates the relationship between the orbital height h, a, i and P_n.

h (km)	a (km)	i (°)	P_n (minutes)
250	6628	96.5	89.6
500	6878	97.4	94.8
750	7128	98.4	100.0
1000	7378	99.5	105.3
1250	7628	100.7	110.7
1500	7878	102.0	116.1
1750	8128	103.4	121.7
2000	8378	104.9	127.4

Supervised classification The process of assigning each pixel in an image to one of a number of classes (ground cover types etc.), by comparing the properties of the pixel with the properties of those pixels (*training data*) known to belong to the various classes. The properties (features) of a pixel can be specified by a vector **x** in N-dimensional feature space. The components of this vector will often be the digital numbers or reflectances in each of N spectral bands, but could also be, for example, radar backscatter coefficients in different polarisation states, texture parameters, or single-band digital numbers in a multi-date composite image.

The simplest type of classification is a **parallelepiped classifier** or **box classifier**. N-dimensional 'boxes' are constructed in feature space, so as to enclose all (or a fixed, high proportion) of the training data for each class. A pixel is assigned to class i if its vector **x** lies within the box defined for that class. If the box edges are parallel to the axes of the feature space, the classification rule takes the form

$$\mathbf{x} \text{ belongs to class } i \text{ if } x_{i1a} < x_1 < x_{i1b} \text{ and } x_{i2a} < x_2 < x_{i2b} \text{ and } \ldots,$$

where x_{ija} and x_{ijb} are respectively the lower and upper limits of the j-component of the box for class i.

The parallelepiped classifier is simple and fast. Its disadvantages are that there may be overlaps between the boxes (in which case a pixel could be assigned to more than one class, with no indication as to which is more likely to be correct), or there may be regions of feature space not enclosed by any of the boxes defined by the training data (in which case a pixel whose vector lay in such a region could not be classified).

More sophisticated classification algorithms use the concept of a **discriminant function** $f_i(\mathbf{x})$, such that the most likely class i to which a pixel belongs is given by that for which $f_i(\mathbf{x})$ is either a minimum or a maximum, depending on the definition of the discriminant function. The simplest discriminant function is the **Euclidean distance**:

$$f_i(\mathbf{x}) = |\mathbf{x} - \langle \mathbf{x} \rangle_i|^{1/2},$$

where $\langle \mathbf{x} \rangle_i$ is the mean vector for the training data for class i. The required class i is that for which this function is minimised. Note that this is equivalent to minimising $f_i^2(\mathbf{x})$, which requires slightly less computation.

Maximum-likelihood classification normally uses the following discriminant function, which is maximised for the most likely class:

$$f_i(\mathbf{x}) = -\ln|\mathbf{C}_i| - (\mathbf{x} - \langle \mathbf{x} \rangle_i)^T \mathbf{C}_i^{-1} (\mathbf{x} - \langle \mathbf{x} \rangle_i).$$

In this expression, $\mathbf{C}_i$ is the *covariance matrix* for the training data from class i, and the superscript T denotes a vector transpose. This discriminant function is derived from Bayes' theorem, assuming equal prior probabilities and that the components of $\mathbf{x}$ for each training area are distributed according to a multivariate normal distribution.

The maximum-likelihood classifier is usually more powerful than the Euclidean distance classifier, but requires more training data in order to generate statistically valid estimates of the covariance matrices $\mathbf{C}_i$. Intermediate between the two is use of the **Mahalanobis distance** as a discriminant function. This is a minimum for the required class, and is defined as:

$$f_i(\mathbf{x}) = \left((\mathbf{x} - \langle \mathbf{x} \rangle_i)^T \mathbf{C}^{-1} (\mathbf{x} - \langle \mathbf{x} \rangle_i) \right)^{1/2},$$

where $\mathbf{C}$ is the covariance matrix calculated for all the training data, or for the whole image.

Thresholding is sometimes applied to image classification using discriminant functions, i.e. the classification on the basis of the maximum (or minimum) value of $f_i(\mathbf{x})$ is accepted only if the value is above (or below) a threshold that represents a minimum acceptable probability. Otherwise, the pixel is not classified.

The number N of dimensions present in the image data may be quite large, resulting in a large processing burden. For this reason, some reduction in the number of dimensions used by the classifier may be desirable. This can be effected after the training stage by calculating the *principal components* or *canonical components* of the data, and using only the first few, or, for example, by combining appropriate bands into *vegetation indices*.

See also *classification, contextual classification, unsupervised classification*.

Supplementary Sensor E See *SSE*.

Surface Composition Mapping Radiometer See *SCMR*.

Surface scattering See *scattering models*.

Swath width The swath of a sensor is the strip of the Earth's surface from which data are collected. The longitudinal extent of the swath is defined by the motion of the platform (satellite) with respect to the surface, although if the sensor does not view symmetrically about the nadir direction the centre-line of the swath will not coincide with the path followed by the sub-satellite point. The swath width is measured perpendicularly to the longitudinal extent of the swath.

Swedish National Space Board The Swedish space agency. See *CEOS*.

Synchronous Meteorological Satellite See *GOES*.

Synthetic aperture radar An imaging *radar* technique, essentially similar to *side-looking radar* (SLR) but providing substantially improved azimuth resolution by more sophisticated processing of the received signal, relying on the motion of the radar with respect to the target. As an *active* remote sensing system, like SLR, synthetic aperture radar (SAR) has the great advantage over visible-wavelength imagers of independence of daylight conditions. Both SLR and SAR can also penetrate through cloud and, to a limited and frequency-dependent extent, through a vegetation canopy.

SAR instruments are sensitive primarily to surface roughness and dielectric properties, and they find applications in monitoring ocean waves and other surface features such as fronts, eddies, oil slicks and ship wakes. They are also used for operational forecasting of *sea ice* conditions. Over land, SAR instruments are used in vegetation monitoring, especially for agriculture and forestry, and for measuring *soil moisture*. SAR also has important applications to the monitoring of *snow, ice sheets* and *glaciers*.

The azimuth resolution of the SLR is limited by the short length of the *antenna* in the along-track direction. In an SAR system, the signals received at the antenna as it is carried along-track by the platform are combined appropriately in order to synthesise the signal that would have been detected by a much longer antenna. This is most easily thought of in terms of Doppler shift. The signal received at the antenna at a given instant originated from a narrow strip of scatterers lying at the same distance (slant range) from the antenna. The azimuthal extent of this strip is defined by the azimuthal extent of the antenna *power-pattern*, and this is the resolution available from a conventional SLR system. However, the frequency of the received signal varies according to the position along the strip from which it originated. Scatterers lying forward of the nadir point will return a signal at a higher frequency than the transmitted signal, because of the Doppler shift due to the fact that the radar's velocity has a component directed towards these scatterers. Similarly, scatterers lying behind the nadir will give a signal Doppler-shifted downwards in frequency. Thus, frequency analysis of the returned signal permits different azimuth coordinates to be resolved. In order to perform this frequency analysis, it is necessary that the received signal is recorded coherently, i.e. both amplitudes and phases must be recorded.

The signal processing required to generate an image from the data received by an SAR system is very complex, and is performed by dedicated SAR processors at ground stations. The image can not be said to exist until this processing has been carried out. The resolution of the image is determined by the processing that is carried out, but some fundamental limits apply. The range resolution is limited by the pulse length and the imaging geometry in exactly the same way as for SLR imaging. The azimuth resolution cannot be finer than $L/2$, where L is the length of the antenna in the along-track direction. It is also limited by the pulse-repetition frequency of the radar, and this requires a compromise between azimuth resolution and **swath width**, in the sense that high azimuth resolution implies a narrow swath.

SAR imagery is subject to the same geometrical distortions as SLR imagery. As a necessarily coherent imaging technique, it is also subject to *speckle*, which imposes a limit on its radiometric resolution. The Doppler processing of the received signal also introduces an *azimuth shift*, in which the image positions of moving scatterers are displaced in the azimuth direction relative to the positions they would have had if they had been stationary.

A further limitation of SAR imagery is caused by the limited number (usually just one or two) of independent bands of data generated by spaceborne SAR systems. Most systems to date have operated at a single frequency, and very few have provided multipolarisation imagery. This can cause significant ambiguities in the interpretation of SAR imagery, which can sometimes be partially removed by the use of multitemporal observations of a particular region. Despite these limitations, SAR imagery is finding an increasingly wide range of applications, mainly as a result of its ability to provide high-resolution imagery even during darkness and through cloud cover. For some applications, notably vegetation monitoring, the sensitivity to the dielectric contrast and the centimetre-scale geometry of the target material is also a direct advantage. The technique of *interferometric SAR* is also finding increasing applications to high-precision topographic mapping and change detection.

The table below summarises the characteristics of the principal spaceborne SAR systems. Note: X = cross-polarised mode, i.e. HV/VH.

Satellite	Launch (year)	Frequency (GHz)	Polarisation	Incidence angle (°)	Spatial res. (m)	Swath width (km)
Seasat	1978	1.28	HH	20	25	100
SIR-A	1981	1.3	HH	38	24	50
SIR-B	1984	1.3	HH	15–60	25	30–60
Kosmos 1870	1987	3.13	HH	25–60	25	20
Almaz 1	1991	3.13	HH	25–60	15	40
ERS-1, -2	1991	5.3	VV	23	30	100
JERS-1	1992	1.28	HH	35	18	75
Mir-1	1993	1.28, 3.28	HH, VV	35	50	50
SIR-C	1994	1.25	HH, VV, X	15–55	40	40–90
SIR-C	1994	5.3	HH, VV, X	20–55	25	40–90
SIR-C	1994	9.6	VV	20–55	30	15–45
Radarsat	1995	5.3	HH	10–60	10–100	45–500
Resurs–O2	1995	1.28	HH	35	100	50
Almaz 1B	1998	8.6	VV	25–51	5	20
Almaz 1B	1998	3.13	HH, VV, X	21–51	5–40	30–170
Almaz 1B	1998	0.43	HH, VV, X	25–51	22–40	120–170
Envisat	1999	5.3	HH, VV	15–45	30–1000	5–400
ALOS	2002	1.28	HH, VV	20–55	10–100	70–250

Tasselled-cap transformation A transformation performed on each pixel of an image, or of a specified region of an image, to yield a set of values x'_i ($i = 1, 2, 3, \ldots$) that are (a) strongly correlated with different vegetation and soil properties, and (b) statistically independent of one another. The tasselled-cap transformation is essentially a *principal components transformation* optimised for vegetated areas. It can be defined as follows:

$$x'_i = \sum_{j=1}^{N} a_{ij} x_j + b_i,$$

where x_j ($j = 1, \ldots, N$) is the digital number of the pixel in band j, a_{ij} are a set of coefficients, and b_i are a set of offsets defined for convenience to ensure that all the x'_i are positive.

The original tasselled-cap transformation (also known as the Kauth–Thomas transform) was defined by Kauth and Thomas for *Landsat MSS* bands 4 to 7, using imagery from agricultural areas in the U.S.A. Taking x_1 as the band 4 DN (in the range 0–127), x_2 as band 5 (0–127), x_3 as band 6 (0–127) and x_4 as band 7 (0–63), the transformation parameters are

$$\mathbf{a} = \begin{pmatrix} 0.433 & 0.632 & 0.586 & 0.264 \\ -0.290 & -0.562 & 0.600 & 0.491 \\ -0.829 & 0.522 & -0.039 & 0.194 \\ 0.223 & 0.012 & -0.543 & 0.810 \end{pmatrix} \quad \mathbf{b} = \begin{pmatrix} 32 \\ 32 \\ 32 \\ 32 \end{pmatrix}.$$

The transformed values are termed 'brightness' (x'_1), associated mainly with variations in the soil background; 'greenness' (x'_2), associated with variations in green vegetation, and containing similar information to the *vegetation index*; 'yellowness' (x'_3), associated with the yellowing of senescent vegetation; and 'nonesuch' (x'_4), possibly related to atmospheric variations.

Analogous transformations have been defined for *Landsat TM* imagery, using bands 1 to 5 and 7. Crist's transformation defines three properties: 'brightness' (coefficients 0.304, 0.279, 0.434, 0.559, 0.508, 0.186); 'greenness' (−0.285, −0.244, −0.544, 0.724, 0.084, −0.180); and 'wetness' (0.151, 0.179, 0.330, 0.341, −0.711, −0.457).

Both of these transformations use predefined coefficients, derived from agricultural areas in the U.S.A. Their suitability for other types of vegetation can not be guaranteed.

TEC See *ionosphere*.

Television Infrared Observation Satellites See *TIROS*.

Temperature and Humidity Infrared Radiometer See *THIR*.

Temperature, land surface Like *sea surface temperature*, the land surface temperature provides an important input to global climate models and a validation of numerical weather predictions. When combined with *precipitation* data it can be used to estimate evaporation rates and hence water fluxes, and to deduce *vegetation* cover. Also on global and regional scales, land surface temperature data can be used for such applications as resource exploration, and monitoring of forest fires and plate-tectonic activity. On local scales, the data can be used to study 'heat islands' in *urban areas*, to predict the occurrence of ground frost, and to identify optimum planting times in agriculture.

Spaceborne remote sensing methods for estimating land surface temperature are currently based on thermal infrared radiometry using data from medium-resolution imagers on *LEO* satellites and (for global applications) low-resolution data from geostationary meteorological satellites. The data require correction for surface *emissivity* and vegetation effects, and also *atmospheric correction*, and the temporal and spatial resolutions that are currently achievable are still rather poor.

Temperature, sea surface See *sea surface temperature*.

Temperature sounding, atmospheric For temperature sounding, radiation emitted by gases that are well-mixed in the atmosphere, such as CO_2 and O_2, is usually measured by nadir-viewing passive infrared and microwave radiometers. This is so that the amount of gas at any level only depends on the atmospheric pressure. The pressure-dependent spectral broadening of the absorption line allows the pressure, and hence height, to be identified. Data from the 15 μm CO_2 band, the 4.3 μm CO_2/N_2O band and the 5 mm O_2 band are used on the *HIRS/MSU* instruments. The infrared data, which have a typical spatial resolution of 10 km (vertically) by 10–100 km (horizontally) and a temperature resolution of a few K, must be screened for *cloud*, with extensively cloud contaminated data being rejected and corrections being applied to partly cloudy data. Microwave sounding instruments can provide substantially lower sensitivity to cloud, but at the expense of poorer spatial resolution.

A number of different methods are available to carry out the inversion necessary to derive temperature. A minimum variance (regression) method can be applied although, as there is no unique solution to the inversion problem, a 'first-guess' atmospheric temperature profile must be applied, either from climatology or a numerical weather prediction system. Alternatively, a physical inversion procedure can be carried out where the forward problem (calculating a radiance at the top of the atmosphere from a temperature profile) is solved using a 'first-guess' profile and the result compared to the satellite measurement. This allows the temperature profile to be altered within an iterative procedure until there is reasonable correspondence between the computed radiance and the satellite measurement.

Atmospheric temperature sounding (profiling) is of fundamental importance in both global climate studies and also in numerical weather forecasting. For the latter application, satellite-based measurements do not currently have sufficient spatial or temperature resolution, and they are combined with surface-based measurements and the predictions of numerical models.

Temporal resolution See *resolution*.

TES (Tropospheric Emission Spectrometer) U.S. thermal infrared spectrometer planned for inclusion on *EOS-Chem* satellites. TES is a Fourier transform spectrometer that can operate as both a limb sounder and as a nadir sounder, and will be used to obtain atmospheric concentrations of water vapour, O_3, CO, SO_2 and other trace gases. Waveband: 2.3 µm to 15.4 µm. Spectral resolution: 750 MHz. Spatial resolution: 2.3 km (vertical – as limb sounder); 50×5 km or 5×0.5 km (horizontal – as nadir sounder). Height range: 0–32 km (as limb sounder).

Tesseral addressing A method of *compression* for digital image data that makes use of spatial homogeneity in the image. A homogeneous (i.e. all pixels having the same value), contiguous region in the image can be specified by a single pixel value and a code (address) that specifies the shape and position of the region. The addresses are hierarchical, such that short addresses represent large areas of the image and long addresses represent small areas. The figure below illustrates one possible hierarchical addressing scheme. The first digit (0, 1, 2 or 3) of an address shows in which quarter of the image a region lies. The second digit shows in which quarter of that quarter a region lies, and so on.

	10	11	
0	120	121	13
	122	123	
2		3	

Using such a scheme, the region illustrated below can be represented by the addresses (002) (003) (02) (03) (112) (200) (201) (210) (211).

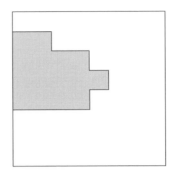

Tesseral addressing can be used for reversible data compression if the full addresses are retained, or for irreversible compression if the addresses are truncated at a point corresponding to a certain degree of spatial averaging. As with *run-length coding*, tesseral addressing is best suited to data with a high degree of spatial homogeneity, for example smoothed or classified imagery.

Texture Although a precise definition of texture is elusive, it can be loosely defined as structure in the spatial variation of the intensity of an image. Recognition of characteristic textures of different ground-cover classes plays an important role in the visual interpretation and classification of images, and attempts to duplicate this process in automated image classification have shown that it can often provide significantly increased classification accuracies over purely multispectral classification.

Since texture is a spatial property, any textural parameter must necessarily be defined by examining variations of the digital number in a finite part of the image. The texture parameter associated with a particular pixel is calculated from the digital numbers in a 'window' centred on that pixel. The choice of window size is important: if it is too small, too few pixels will be available to give a statistically useful measure of texture, while if it is too large the spatial resolution of the 'texture image' will be degraded.

The simplest quantitative measure of texture is the variance of the digital numbers in the window, but this gives no information on the spatial frequency of the variation. More sophisticated methods involve calculating and then parametrising the *Fourier transform, Hadamard transform* or *semivariogram* of the data within the window. Most use is, however, probably made of the *grey level co-occurrence matrix*.

Thematic map See *classification*.

Thematic mapper See *TM*.

Thermal inertia A measure of a material's resistance to a change in temperature in response to a temperature change in its surroundings. The thermal inertia P

is defined as

$$P = \sqrt{K\rho c},$$

where K is the thermal conductivity, ρ is the density and c is the specific heat capacity. It is thus related to the thermal diffusivity (see *diffusion equation*). If the heat flux into the surface of a uniform material of thermal inertia P varies sinusoidally at angular frequency ω and with amplitude F, the surface temperature oscillations will lag the flux oscillations by an eighth of a cycle and will have amplitude $F/P\sqrt{\omega}$. For example, dry sandy soils have a low thermal inertia and reach relatively high temperatures during the day and relatively low temperatures at night, while wet clay soils, with a higher thermal inertia, are more resistant to temperature change and will have a smaller diurnal temperature range.

The apparent thermal inertia is, to a first approximation, relatively simple to estimate from pairs of thermal infrared images recorded during the day and night. Vegetation, water and damp soils have similar low to medium thermal inertias and so apparent thermal inertia images of temperate agricultural scenes offer little differentiation of land cover. Apparent thermal inertia images have proved to be of particular value for the differentiation of unvegetated surfaces in arid and semi-arid regions, usually as a prelude to geological mapping, and in the identification of geological units and of fault and fold zones (see *geology*). Materials with a low thermal inertia include pumice, ignimbrite, sand and shale; sandstone, marble, basalt and granite have intermediate thermal inertia; and serpentinite, pterodinite, dolomite and quartzite have high thermal inertia.

Typical values of the density ρ, specific heat capacity c, thermal conductivity K, thermal inertia P and thermal diffusivity Γ for various materials are given in the table below.

Material	ρ (10^3 kg m^{-3})	c (kJ kg^{-1} K^{-1})	K (W m^{-1} K^{-1})	P (10^3 J s$^{-1/2}$ m^{-2} K^{-1})	Γ (10^{-6} m^2 s^{-1})
Geological materials – low inertia					
clay (moist)	1.7	1.5	1.3	1.8	0.5
gravel	2.1	0.8	1.3	1.5	0.8
limestone	2.5	0.7	0.9	1.3	0.5
obsidian	2.4	0.71	1.25	1.5	0.7
pumice (loose)	1	0.7	0.3	0.5	0.4
sand (dry)	1.6	0.8	0.4	0.7	0.3
shale	2.3	0.7	1.9	1.7	1.2
soil (sandy)	1.8	1	0.6	1.0	0.3
tuff	1.8	0.84	1.17	1.3	0.8
Geological materials – intermediate inertia					
basalt	2.6	0.9	2.1	2.2	0.9
gabbro	3	0.71	2.5	2.3	1.2
granite	2.7	0.8	3	2.5	1.4
gravel (sandy)	2.1	0.8	2.5	2.0	1.5
marble	2.7	0.9	2.5	2.5	1.0
rhyolite	2.5	0.67	2.3	2.0	1.4
sandstone	2.5	0.8	3.8	2.8	1.9
slate	2.8	0.71	2.1	2.0	1.1
syenite	2.2	0.96	2.5	2.3	1.2

Material	ρ (10^3 kg m^{-3})	c $(\text{kJ kg}^{-1}\text{ K}^{-1})$	K $(\text{W m}^{-1}\text{ K}^{-1})$	P $(10^3 \text{ J s}^{-1/2}\text{ m}^{-2}\text{ K}^{-1})$	Γ $(10^{-6} \text{ m}^2\text{ s}^{-1})$
Geological materials – high inertia					
dolomite	2.6	0.75	2	2.0	1.0
peridotite	3.2	0.84	4.6	3.5	1.7
quartz	2.6	0.7	9	4.0	4.9
quartzite	2.7	0.7	5	3.1	2.6
serpentine	2.4	0.96	2.83	2.6	1.2
Other materials					
concrete	2.4	3.4	0.1	0.9	0.01
glass	2.3	0.6	0.8	1.1	0.6
ice	0.9	2.1	2.3	2.1	1.2
metals	5–10	0.2–0.5	20–100	5–20	5–100
water	1	4.2	0.56	1.5	0.1
wood	0.7	1.2	0.15	0.4	0.2

Thermal infrared See *infrared*.

Thermal infrared radiometry Determination of the Earth's surface (land, water or ice) temperature by detecting thermal infrared radiation. Calibrated data are usually in the form of *brightness temperatures*, and may include *atmospheric correction*. In order to derive the surface temperature it is necessary to know the *emissivity* of the target material.

Thermosphere The uppermost layer of the *atmosphere*, extending from the mesopause, at an altitude of about 90 km, to 400 km or more.

THIR (Temperature and Humidity Infrared Radiometer) U.S. thermal infrared imaging radiometer, carried on *Nimbus-4* to *-7*. Wavebands: 6.5–7.0, 10.5–12.5 μm. Spatial resolution: 23 km, 8 km, respectively, at nadir. Swath width: 3000 km.

Tides Periodic variations in the height of the ocean surface as a result of the gravitational attraction of the Sun and Moon. The principal tidal modes are listed below (h = hour; d = day = 24 h; w = week = 7 d):

Symbol	Name	Period
K_2	semidiurnal declination lunisolar	11.967 h
S_2	semidiurnal principal solar	12.000 h
M_2	semidiurnal principal lunar	12.421 h
N_2	semidiurnal elliptical lunar	12.658 h
K_1	diurnal declination lunisolar	23.935 h
P_1	diurnal principal solar	24.066 h
O_1	diurnal principal lunar	25.819 h
Q_1	diurnal elliptical lunar	26.868 h
M_f	fortnightly lunar	13.661 d
M_m	monthly lunar	27.555 d
S_{sa}	semiannual solar	26.090 w

Tidal variations can be studied using *radar altimeter* observations, though caution must be exercised to ensure that the tidal frequency is not reduced to an excessively low frequency by *aliasing*.

TIROS A series of 10 early U.S. satellites, operated by *NOAA* and functioning from April 1960 (launch of TIROS-1) to April 1966 (decommissioning of TIROS-10). Objectives: Experimental and operational meteorological observations. Orbit: Nominally circular *LEO* (eccentricities below about 0.2) except TIROS-5 ($e = 0.33$) and TIROS-9 ($e = 0.64$). TIROS-9 and -10 were *Sun-synchronous*. Minimum altitudes between 680 and 850 km; maximum altitudes between 740 and 970 km except TIROS-5 (1120 km) and TIROS-9 (2970 km). Period 97 to 101 minutes (TIROS-9: 119 minutes); inclination 48° (TIROS-1 to -4), 58° (TIROS-5 to -8), 96° (TIROS-9), 98° (TIROS-10). Principal instruments: *APT* (TIROS-8), *SR* (TIROS-2, -3, -4 and -7), *Vidicon*.

TIROS-9 was the first Sun-synchronous satellite.

TIROS N U.S. satellite, operated by *NOAA*, launched October 1978. Objectives: Operational meteorological satellite. Orbit: Circular *Sun-synchronous LEO* at 850 km altitude. Period 102.3 minutes; inclination 98.9°. Principal instruments: *AVHRR, HIRS-2, MSU, SSU*. The satellite also carried the *Argos* data collection system.

TIROS Operational Vertical Sounder See *TOVS*.

TK-350 Russian *photographic system* on *Kosmos* satellites. Waveband: 0.49–0.59 μm, Spatial resolution: 8 m. Image size: 170 × 265 km.

URL: http://www.eurimage.it/Products/KVR_1000.html

TM (Thematic Mapper) U.S. optical/near infrared/thermal infrared mechanically scanned imaging radiometer, carried on *Landsat-4* and *-5* satellites. Wavebands: 1: 0.45–0.52 μm; 2: 0.52–0.60 μm; 3: 0.60–0.69 μm; 4: 0.76–0.90 μm; 5: 1.55–1.75 μm; 6: 10.4–12.5 μm; 7: 2.08–2.35 μm. Spatial resolution: 30 m (120 m for band 6). Swath width: 185 km.

URL: http://edcwww.cr.usgs.gov/glis/hyper/guide/landsat_tm

TMI (TRMM Microwave Imager) U.S. scanning passive microwave radiometer, carried on *TRMM* satellite. Frequencies: 10.7, 19.4, 22, 37, 88 GHz. Polarisation: H and V. Spatial resolution: 45 km at 10.7 GHz to 4.4 km at 88 GHz. Swath width: 760 km.

The main purpose of the TMI is to derive vertically integrated rainfall rates over ocean areas.

URL: http://hdsn.eoc.nasda.go.jp/guide/guide/satellite/sendata/tmi_e.html

TMR (Topex Microwave Radiometer) U.S. nadir-viewing passive microwave radiometer carried on *Topex-Poseidon* and planned for inclusion on *EOS-Alt*. Its purpose is to measure the total atmospheric water vapour content to provide correction (to ±5 cm) for radar altimeter measurements. Frequencies: 18, 21, 37 GHz. Spatial resolution: 30 km.

URL: http://www-aviso.cls.cnes.fr/English/TOPEX_POSEIDON/More_On_Payload.html

TOMS (Total Ozone Mapping Spectrometer) U.S. scanning ultraviolet filter spectrometer for measuring total atmospheric ozone and SO2 content, carried on *Meteor-3*, *Nimbus-7* and *ADEOS* and planned for inclusion on *ADEOS II* and *TOMS-EP*. Wavebands: 309, 313, 318, 322, 331, 360 nm. Spectral resolution: 1 nm. Spatial resolution: 50 km at nadir (from 800 km). Swath width: 2300 km from 800 km (102° scan across-track in 3° steps).
Operation of the Meteor-3 TOMS instrument ceased in December 1994.

URL: http://jwocky.gsfc.nasa.gov/

TOMS-EP (TOMS Earth Probe) U.S. satellite, operated by *NASA*, launched in July 1996. The nominal lifetime is 2 years. Objectives: Global daily monitoring of atmospheric ozone concentrations. Provision of data continuity with *Nimbus-7* and *Meteor-3* missions. Orbit: Circular *Sun synchronous LEO* at 955 km altitude. Period 104 minutes; inclination 99.3°. Principal instruments: *TOMS*.

URL: http://jwocky.gsfc.nasa.gov/

Topex Microwave Radiometer See *TMR*.

Topex-Poseidon U.S.–French satellite, operated by *NASA* and *CNES*, launched in August 1992 with a nominal lifetime of 3 to 5 years. Objectives: Dedicated altimetry mission for ocean topography, geoid measurement etc. Orbit: Circular *LEO* at 1334 km altitude. Period 112 minutes; inclination 66°. Principal instruments: *ALT*, *SSALT*, *TMR*. The satellite also carries a laser retroreflector array, a radio frequency Doppler location package (*DORIS*) and a *GPS* receiver for precise (5 cm) determination of position.
A follow-on mission, *Jason-1*, is planned as part of the *EOS* programme.

URL: http://topex-www.jpl.nasa.gov/

Topographic mapping Topographic data over the land surface are required for a wide range of applications, including civil planning and development, determination of the susceptibility of coastal areas to flooding, modelling of hydrological drainage, investigation of erosion, monitoring of earthquakes, volcanoes and landslides.
The surface topography of relatively small areas of the Earth's land surface can be measured using *stereophotography* , stereo *radargrammetry*, or *interferometric SAR* methods. However, for regional and global-scale topographic mapping, the topography of the land, ice and sea floor have been measured by

satellite altimeters. For the sea floor, the *Seasat, Geosat, ERS, Topex-Poseidon* and Russian *Geoik* satellites have used *radar altimeters* to measure very accurately the range between the satellite and the sea surface. If the distance of the satellite from a reference surface can be accurately determined then the elevation of the sea surface from that reference surface will be accurately known. Within a restrictive wavelength range (about 15–160 km), the elevations of the sea surface, with respect to that reference surface, reflect variations in sea floor topography. This is because features on the sea floor exert gravitational attractions on the water above and cause a slight increase or decrease in the surface of the ocean, through isostatic compensation. In this way topographic features of all the sea floors, including those of the Arctic Ocean, have been revealed during the last 15 years. The technique of satellite radar altimetry has also revealed the topography of land ice surfaces but has been less successful over land surfaces, because the backscattered radar signals do not follow the topographic variations over the relatively large radar signal *footprint* on the ground.

Over land surfaces satellite *laser altimeters* are beginning to be used for surface topography. For several years laser-based satellite instruments have been used to measure the configuration of cloud tops and to study the vertical distribution of aerosols and the planetary boundary characteristics of the atmosphere. Starting in early 1996 the first land topography-oriented satellite laser altimeter system was flown in the Space Shuttle and future flights of this system are anticipated in 1998 and beyond. This system provides a complete record of the range to the surface as well as a measure of the within-footprint surface height distribution. Sub-metre vertical precision measurements have been achieved and when combined with metre-level orbit accuracy, a highly accurate dataset of Earth surface ground elevation points can be generated from these data.

All satellite altimeters, whether radar or laser, require precise radial orbit knowledge (knowledge of the *geoid*). The satellite's coordinates are determined by surface tracking and the geoid is expressed by a mathematical model of the Earth's gravity field. This model has been developed by research programmes that use laser ranging on special satellites with laser reflectors and on other satellites.

See also *sea surface topography*.

Total electron content See *ionosphere*.

Total Ozone Mapping Spectrometer See *TOMS*.

Total Ozone Mapping Spectrometer – Earth Probe See *TOMS-EP*.

TOVS (TIROS Operational Vertical Sounder) Package of three instruments, for measurement of temperature, water vapour, ozone and carbon dioxide profiles, and cloud distributions, carried by *NOAA-9* onwards. TOVS consists of the *HIRS-2*, *SSU* and *MSU* (*AMSU-A/MHS* after 1993) instruments.

TPFO (Topex-Poseidon Follow-On) See *Jason-1*.

Trace gases, atmospheric See *chemistry, atmospheric*.

Training data *Pixels* in an image, selected to be representative of a particular class (e.g. land-cover type) in the entire image. The training data are used to 'train' the *classification* algorithm, i.e. to derive the rules on which the classification will be based. It is important to ensure that: (i) the training data are representative of all the data for the class in question; (ii) there are sufficient pixels to ensure that their statistical properties are well determined; (iii) the statistics of the training data are compatible with the model assumed by the classification algorithm.

Transformed divergence See *separability*.

Transformed vegetation index See *vegetation index*.

Transponder See *radar transponder, laser retroreflector*.

Transverse Mercator Projection See *map projection*.

Travers Russian dual-frequency *synthetic aperture radar*, carried on Priroda module of *Mir-1* space station. Frequencies: L band (1.3 GHz) and S band (3.3 GHz). Polarisation: VV and HH. Incidence angle: 35°. Spatial resolution: Range direction 100 m; azimuth direction 20 m (for data processing on the ground), 150 m (for data processing on board). Swath width: 50 km.

TRMM (Tropical Rainfall Measuring Mission) U.S.-Japanese satellite, operated by *NASA* and *NASDA* and launched November 1997 with a nominal lifetime of 3 years. TRMM is part of the *EOS* programme. Objectives: Measurement of precipitation and evaporation in tropical areas, for atmospheric circulation and energy budget studies. Orbit: Circular *LEO* at 350 km altitude. Period 92 minutes; inclination 35°. Principal instruments: *CERES, LIS, PR, TMI, VIRS*.

URL: http://www.nasda.go.jp/
http://trmm.gsfc.nasa.gov/

TRMM Microwave Imager See *TMI*.

Tropopause The boundary between the *troposphere* and the *stratosphere*.

Troposphere The lowest layer of the *atmosphere*, extending from the Earth's surface to the tropopause (typically 8 km at the poles, 18 km at the equator). The temperature gradient in the troposphere is negative; it has a high water vapour content, is turbulent, and is the most significant layer of the atmosphere in terms of weather.

Tropospheric delay Pulses of electromagnetic radiation propagate through the atmosphere at a speed less than the speed of light *in vacuo*. The pulse propagation time is determined by the group velocity, which will differ from the phase velocity if the medium is dispersive. The group velocity also varies with height, due to variations in atmospheric density, and most of the delay occurs in the troposphere. The delay can be expressed as the extra time taken to traverse a path through the atmosphere compared with the time taken to cover the same path *in vacuo*. This time is then converted to a distance by multiplying by c, giving the range error.

The atmosphere can be divided into two components: (1) the dry atmosphere (principally nitrogen, oxygen and carbon dioxide); (2) water vapour. The range error in each case depends on the integrated number density of molecules along the path. For the **dry atmosphere** component, and for a vertical path through the entire atmosphere, this term is proportional to the pressure at the bottom of the atmosphere. The figures given below therefore relate to a partial pressure of dry air of one standard atmosphere (101 325 Pa) at the bottom of the atmosphere. For a path that does not traverse the entire atmosphere, and that makes an angle θ with the vertical, the dry atmosphere range error should be multiplied by

$$\frac{p_{\text{bottom}} - p_{\text{top}}}{\cos \theta},$$

where p_{bottom} and p_{top} are the partial pressures of dry air at the bottom and top, respectively, of the path, both measured in atmospheres. This formula is not valid for very large values of θ.

For the **water vapour** component, the integrated number density of molecules is normally expressed as the thickness of the layer of water that would result if the water vapour were precipitated. The range errors below are therefore expressed as metres per metre of precipitable water. For a path making an angle θ with the vertical, they should again be divided by $\cos \theta$.

Ultraviolet to near infrared (dispersive)

λ (μm)	Dry atmosphere (metres per atmosphere)	Water vapour (metres per metre of precipitable water)
0.2	3.99	1.57
0.3	2.81	0.83
0.4	2.56	0.59
0.5	2.46	0.49
0.6	2.41	0.43
0.7	2.38	0.40
0.8	2.36	0.37
0.9	2.34	0.36
1.0	2.33	0.35
1.5	2.31	0.32
2.0	2.30	0.31

Radio (non-dispersive)

Dry air: 2.33 metres per atmosphere; water vapour 7.1 metres per metre of precipitable water.

See also *ionosphere*.

Tropospheric Emission Spectrometer See *TES*.

Tropospheric Rainfall Measuring Mission See *TRMM*.

Turbidity Turbidity is the attenuation of the direct beam solar radiation that occurs in a cloud-free atmosphere as a result of scattering. It is mainly determined by the optical depth due to aerosols, which can be dust, smoke particles or other suspended material in the *troposphere* or *stratosphere*. The turbidity is an important parameter in many short-wave radiative transfer calculations used for studies into solar energy and agrometeorology. Turbidity is mainly estimated via the attenuation of the solar radiation using surface-based pyrheliometers. The turbidity factor is defined by

$$I = I_0 \exp(-T_L \tau_s m),$$

where I is the normal incidence direct irradiance at the surface, I_0 the extraterrestrial direct irradiance, τ_s the broad-band Rayleigh scattering optical depth, m the air mass number and T_L is the turbidity factor.

The effects of turbidity are apparent in some visible-wavelength satellite imagery, especially where the aerosol has a high albedo and is over a low albedo surface, such as the ocean. The ocean area to the west of the Sahara desert occasionally has a brighter appearance in imagery as Saharan dust is blown out via an easterly wind. It can be distinguished from thin cirrus by its persistence over several days. Volcanic dust can be injected into the stratosphere where it can persist for long periods, giving reduced solar radiation at the Earth's surface and affecting the derivation of sea and land surface temperatures from remotely sensed data.

Turbulence, atmospheric Irregular, turbulent flow, characterised by a range of motions on different timescales and space scales (eddies), is the most commonly encountered flow in the atmosphere. Only very close to the Earth's surface is the flow smooth and laminar without mixing of fluid elements with their surroundings. In the *boundary layer* of the atmosphere (the layer where the fluid motion is affected by the surface) turbulent flow is responsible for most of the vertical transfer of heat, momentum and mass. The inclination of the atmosphere to vertical overturning is dependent on the stability of the atmosphere, which is a function of the vertical profile of temperature. The stability can sometimes be inferred from satellite imagery by the form of the clouds. Shallow cumulus clouds indicate unstable conditions near to the surface, while cumulonimbus clouds suggest instability through a significant depth of the troposphere.

TV (TV optical instrument) Russian *vidicon* visible-wavelength imaging radiometer, carried on *Meteor-1* and *-2*, and *Meteor-Priroda*, satellites. Waveband: 0.4–0.8 µm. Spatial resolution: 1.25 km at nadir. Swath width: 1000 km.

TVI (Transformed Vegetation Index) See *vegetation index*.

UARS (Upper Atmosphere Research Satellite) U.S. satellite, operated by *NASA*, launched in September 1991. Objectives: Measurement of the energy budget and of trace gas and temperature profiles in the upper atmosphere. Orbit: Circular *LEO* at 585 km altitude. Period 97 minutes; inclination 57°. Principal instruments: *CLAES*, *HALOE*, *HRDI*, *ISAMS*, *MLS*, *WINDII*. The satellite also carries solar–terrestrial physics packages.

> URL: http://umpgal.gsfc.nasa.gov/uars-science.html

UHF radiometer Russian thermal infrared and passive microwave imaging radiometer, carried on *Almaz-1* satellite.
Infrared unit: Wavebands: 11, 12, 13.7 µm. Spatial resolution: 5 km. Swath width: 500 km.
Microwave unit: Frequencies: 6.0, 37.5 GHz. Spatial resolution: 10–30 km. Swath width: 500 km.
  The principal function of the instrument is to collect sea and land surface temperature data.

Ultraviolet The region of the *electromagnetic spectrum* with wavelengths between 10 nm and 380 nm. Ultraviolet radiation can cause *fluorescence* of some minerals and vegetation. Solar radiation contains appreciable amounts of ultraviolet radiation, although this is significantly attenuated by scattering and absorption in the atmosphere. Observation of backscattered solar ultraviolet radiation forms the basis of one method of detecting atmospheric *ozone*.

Unpolarised See *polarisation*.

Unsupervised classification The process of assigning each pixel in an image to one of a number of distinguishable information classes, using *clustering* techniques. Unlike *supervised classification*, unsupervised classification does not ensure that the information classes must correspond to the ground-cover classes required by the user; it does, however, maximise the distinguishability of the classes. See *classification, hybrid classification*.

Upper Atmosphere Research Satellite See *UARS*.

UPS (Universal Polar Stereographic) See *map projection*.

Urban areas Delineation of urban areas can be performed using a variety of types of spaceborne imagery. If data of sufficiently high resolution are available, characteristic structures such as houses, city blocks, road networks etc. can be identified and used to map the extent of the area, and detail within it. However, urban areas can also be mapped using lower resolution imagery. Urban areas generally show significantly different spectral reflectance properties from the surrounding terrain in daytime optical/near infrared imagery, and night-time imagery of sufficient radiometric sensitivity can reveal built-up areas through characteristic patterns of street lighting (see *population estimation*). Thermal infrared imagery shows the 'urban heat island' effect whereby urban areas can be up to a few degrees warmer than the surrounding terrain. Built-up areas also give a characteristically high response in synthetic aperture radar imagery, principally as a result of the large density of horizontal and vertical surfaces meeting at right angles and giving strong dihedral or trihedral scattering responses.

User's accuracy See *error matrix*.

UTM (Universal Transverse Mercator) See *map projection*.

Variogram See *semivariogram*.

VAS (VISSR and Atmospheric Sounder) U.S. optical/infrared spin-scan imaging radiometer, carried on *GOES*-4 to -7. Wavebands: 0.55–0.75 μm (8 bands), 3.9–14.7 μm (12 bands). Spatial resolution: 0.9 km (optical), 7–14 km (infrared). Field of view: Full Earth disc as seen from *geostationary* orbit. Scan time: 30 minutes (full disc).
VAS is an extension of the *VISSR* carried by earlier GOES satellites.

V band Subdivision of the microwave region of the *electromagnetic spectrum*, covering the frequency range 46 to 56 GHz (wavelengths 5.4 to 6.5 mm).

Vector format The representation of a linear feature, such as a contour or the boundary between two regions, on a map or image by a series of coordinates representing positions along the line. Vector format data usually arise from digitising maps etc. Digital image processing methods normally operate on *raster format* data, so a vector-to-raster transformation must first be performed on vector format data.

Vegetation French optical/infrared imaging radiometer planned for inclusion on *SPOT-4* satellite, for global vegetation monitoring. Wavebands: 0.43–0.47, 0.61–0.68, 0.78–0.89, 1.58–1.75 μm. Spatial resolution: 1.2 km at nadir. Swath width: 2200 km.

Vegetation, electromagnetic properties
Optical/near infrared region
Healthy vegetation has a generally characteristic spectral reflectance in the optical and near infrared regions of the electromagnetic spectrum. Below about 0.7 μm, the reflectance is dominated by pigment absorption, giving a low reflectance (typically below 10 to 15%). The reflectance rises sharply, to typically 45–65%, between about 0.65 μm and 0.75 μm (the *red edge*), declining gradually with increasing wavelength. Superimposed on this declining reflectance are minima near 1.45 μm and 1.9 μm due to absorption by water. The figure below illustrates the typical spectral variation of healthy vegetation. Variations in this general trend are often species-specific, but can also be due to factors such as structure, senescence, water content and *soil* type. See also *leaf, vegetation index, geological applications*.

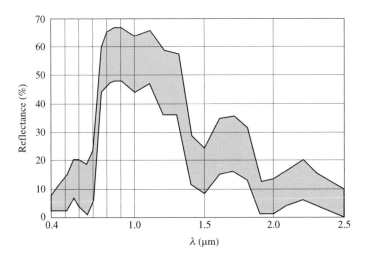

Thermal infrared region

Typical emissivities are listed below:

Material	3–5 µm	8–14 µm
bark	0.87–0.90	0.94–0.97
coniferous twigs, green	0.96	0.97
grass, dry	0.82	0.88
leaves, green	0.86–0.94	0.90–0.96

Vegetation index A mathematical operation performed on the reflectances measured in two or more spectral bands of an optical/near infrared image to yield a parameter that is correlated with the amount of vegetation present in the pixel. The basis of the vegetation index is the high reflectance of leaves in the near infrared due to multiple scattering in the mesophyll, together with visible-wavelength absorption due to plant pigments, principally chlorophyll. Reflectance measurements in the optical and near infrared bands are thus strongly correlated with the fraction of photosynthetically active radiation absorbed by the plant material, and hence with the rate of primary production, and they consequently find an important application in *vegetation mapping*.

Many vegetation indices have been proposed, mostly defined in terms of the reflectance R_r in the red part of the optical spectrum and the reflectance R_i in the near infrared. The precise definition of the spectral bands depends on the sensor to which the vegetation index is applied, but the red reflectance is generally measured in a band between approximately 0.6 and 0.7 µm (e.g. *Landsat MSS* band 5), and the infrared reflectance in a band from 0.8 to 1.0 or 1.1 µm (e.g. MSS band 7). The main vegetation indices are the ratio vegetation index (RVI), normalised difference vegetation index (NDVI), transformed vegetation index (TVI), perpendicular vegetation index (PVI), weighted difference vegetation index (WDVI) and soil-adjusted vegetation index (SAVI). These are defined below:

$$\text{RVI} = \frac{R_i}{R_r},$$

$$\text{NDVI} = \frac{R_i - R_r}{R_i + R_r},$$

$$\text{TVI} = \sqrt{\text{NDVI} + 0.5},$$

$$\text{PVI} = \frac{R_i - \gamma R_r}{\sqrt{1 + \gamma^2}}$$

$$\text{WDVI} = R_i - \gamma R_r,$$

$$\text{SAVI} = \frac{R_i - R_r}{R_i + R_r + L}(1 + L).$$

In these expressions, γ is the slope of the 'soil line', i.e. the value (assumed constant) of dR_i/dR_r for bare soil, and L is a parameter which is adjusted to take account of the density of vegetation. It typically takes a value of 0.5.

See also *tasselled-cap transformation*.

Vegetation mapping The distribution and condition of vegetation on the Earth's land surface is of wide significance at all spatial scales. Vegetation distribution affects the global climate through its *albedo* and its role in modulating biogeochemical fluxes, particularly of water and of carbon. In *agriculture*, vegetation monitoring has an obvious role in identifying and monitoring crops, estimating yields and optimal harvesting times, in the management of irrigation, and in the control of pests. Vegetation mapping also finds many applications in *forestry*.

Vegetation mapping is usually performed using optical/near infrared image data, making use of the characteristic spectral reflectance properties of green vegetation (see *vegetation index*). Local or regional satellite vegetation mapping normally uses high-resolution *multispectral imager* data, for example from *Landsat* and *SPOT* satellite sensors. Maps of Global Vegetation Index (GVI – essentially equivalent to NDVI) are generated from *AVHRR* data, and have found major use in studying global vegetation dynamics. The strong correlation between vegetation index and primary productivity means that satellite vegetation index data can be used to estimate photosynthetic activity, and hence regional and temporal variations in carbon dioxide concentrations. Vegetation index data can also be used for regional applications such as the study of deforestation and, when integrated with rainfall data, the estimation of inherent productivity.

Vertical polarisation A term used in side-looking radar (*side-looking radar* or *synthetic aperture radar*) to describe linearly polarised radiation in which the electric field lies in the plane containing the radiation propagation direction and the local vertical. Compare *horizontal polarisation*.

Vertical Temperature Profile Radiometer See *VTPR*.

Very High Resolution Radiometer See *VHRR*.

Very High Resolution Scanning Radiometer See *VHRSR*.

VH-polarisation A term used in *radar*, to describe a signal that is transmitted in *vertical polarisation* and received in *horizontal polarisation*. This is a *cross-polarised* mode of operation.

VHRR (Very High Resolution Radiometer) Generic name for optical/infrared imaging radiometers on several satellites:

INSAT-1: Spin-scan instrument. Wavebands: 0.55–0.75, 10.5–12.5 μm. Spatial resolution: 2.75 km (optical), 11 km (infrared). Field of view: Full Earth disc as seen from *geostationary* orbit. Scan time: 30 minutes (full disc).

INSAT-2: Spin-scan instrument. Wavebands: 0.55–0.75, 10.5–12.5 μm. Spatial resolution: 2.0 km (optical), 8 km (infrared). Field of view: Full Earth disc as seen from *geostationary* orbit. Scan time: 30 minutes (full disc).

NOAA-2 to -5: Mechanically scanned instrument, forerunner of *AVHRR*. Wavebands: 0.60–0.70, 10.5–12.5 μm. Spatial resolution: 1 km at nadir. Swath width: 2600 km.

VHRSR (Very High Resolution Scanning Radiometer) Chinese optical/infrared imaging radiometer, carried on *FY-1B* satellite. Wavebands: 0.48–0.53, 0.53–0.58, 0.58–0.68, 0.73–1.10, 10.5–12.5 μm. Spatial resolution: 1.1 km at nadir. Swath width: 3200 km.
   The function of the instrument is similar to the *AVHRR* and, as with the AVHRR, data can be downloaded in HRPT and APT formats.

Vidicon
1. A type of television camera, and the earliest *electro-optical sensor* used for visible-wavelength remote sensing from space. In operation, an electron beam first coats a photoconductive plate with electrons. A shutter is then opened, and incident light is focussed on to the plate. This leaves a distribution of charge corresponding to the light intensity, which is read off by electromagnetically scanning the electron beam across the plate. For examples, see *APT, AVCS, IDCS, KL-103W, MR-2000M, MR-900B, RBV, STR, TV*.
2. Vidicon sensor carried by the *TIROS* satellites. Waveband: 0.45–0.65 μm. Spatial resolution: 2 km. Swath width: 1000 km.

VIRS (Visible and Infrared Scanner) U.S. optical/infrared mechanically scanned imaging radiometer, carried on *TRMM* satellite. Wavebands: 0.63, 1.6, 3.8, 10.8, 12.0 μm. Spatial resolution: 2 km. Swath width: 720 km.
   The principal application of the instrument will be to the characterisation of clouds.

Visible and Infrared Scanner See *VIRS*.

Visible and Infrared Scanning Radiometer See *VIRSR*.

Visible and Infrared Spin-Scan Radiometer See *VISSR*.

Visible and Thermal Infrared Radiometer See *VTIR*.

Visible-wavelength radiation (Optical radiation) The region of the *electro-magnetic spectrum* to which the human eye is sensitive, corresponding to the range of wavelengths from 0.38 μm to 0.78 μm. This range can be broadly divided into three colours: red (0.38 to 0.5 μm), green (0.5 to 0.6 μm) and blue (0.6 to 0.78 μm), although the human eye can resolve a much larger number of colours on the basis of the spectral composition. Very many remote sensing systems operate in the optical region of the electromagnetic spectrum, detecting the visible component of the incoming solar radiation that is reflected from the Earth.

VISSR (Visible and Infrared Spin-Scan Radiometer) Generic name for optical/infrared imaging radiometers carried on several *geostationary* satellites.

GMS: Wavebands: 0.5–0.75 μm, 10.5–12.5 μm. Spatial resolution: 1.25 km (optical), 5 km (infrared). Field of view: Full Earth disc as seen from geostationary orbit. Scan time: 30 minutes (full disc).

GOES: Wavebands: 0.5–0.7 μm, 10.5–12.5 μm. Spatial resolution: 1.25 km (optical), 5 km (infrared). Field of view: Full Earth disc as seen from geostationary orbit. Scan time: 30 minutes (full disc). See *VAS*.

Meteosat: Wavebands: 0.5–0.9 μm, 5.7–7.1 μm, 10.5–12.5 μm. Spatial resolution: 2.5 km (optical), 5 km (infrared). Field of view: Full Earth disc as seen from geostationary orbit. Scan time: 30 minutes (full disc). Also known as MSR (Multispectral Radiometer) and MVIRI.

VISSR Atmospheric Sounder See *VAS*.

Volcanoes Volcanoes represent windows into the Earth's interior based on the petrology and geochemistry of the erupted magmas. Volcanoes also play an important role in affecting weather and short-term climate on a local to hemispheric scale, while the locations of volcanoes provide important information on the regional tectonics of an area. Volcanoes also constitute a significant natural hazard, not only to people and property on the ground (which are affected by lava flows, mudflows and volcanic flows called 'pyroclastic flows') but also in the air. Many valuable observations of volcanoes and volcanic eruptions can be obtained through remote sensing, which has the advantage not only of making observations at a scale not available to the field vulcanologist, but also of providing data from a distance that does not place the observer at personal risk.

Exciting new observations of volcanoes have been obtained using *SIR-C/X-SAR*, *ERS-1* and *ERS-2* radar images, which show the potential of *interferometric SAR* to measure surface topography at the metre-scale and surface deformation at the centimetre-scale over hundreds of square kilometres. Radar imaging has also provided the first synoptic view of certain volcanic regions such as the Aleutians, where frequent cloud cover, adverse weather on the ground, and long polar nights make the polar regions almost impossible to map by conventional means.

AVHRR and *GOES* weather satellite images provide synoptic views of the world with a sufficiently high spatial resolution to be used for the detection and tracking of eruption plumes. Plume temperature, abundance of silicate material, and mean aerosol particle size can also be determined. Such information is particularly important for the assessment of hazards to air traffic provided that the information is made available to airlines within a few tens of minutes of data acquisition.

Landsat TM data have been used since 1984 to study the thermal properties of volcanic craters and lava flows. Long-term thermal variations have been used to interpret changes in the level of activity of volcanoes in Chile and Sicily. Detailed analyses of the thermal flux from moving lava flows show that the energy distribution in a flow can be used to interpret the chronology of active flows, the location of active lava tubes, and the hazard potential associated with an eruption. The assessment of hazard potential is enhanced through the use of *SPOT* panchromatic and multispectral data, and photographs taken from the *Space Shuttle*, that enable many volcanoes in infrequently studied parts of the world to be mapped in detail.

Numerous techniques are under development for the analysis of volcanic gases from space. Ultraviolet measurements made by the *TOMS* instrument have provided an inventory of all major eruptions since 1978 that have injected SO_2 into the stratosphere. In some cases, TOMS observations have provided the first reports of new eruptions in Africa and the Galapagos Islands. The *MLS* instrument on *UARS*, and the *SAGE-II* instruments, have also tracked volcanic clouds over many months. *HIRS-2* data have also been used to detect SO_2 from eruptions of Mount St Helens volcano.

Volkhov See *infrared radiometers*.

Volume scattering See *scattering models*.

VSAR (Variable off-nadir angle SAR) Japanese *synthetic aperture radar* planned for inclusion on *ALOS*. Frequency: L band. Polarisation: HH, VV. Incidence angle: 20° to 55°. Spatial resolution: 10 m (high-resolution mode); 100 m (Scan-SAR mode). Swath width: 70 km (250 km in ScanSAR mode).

VTIR (Visible and Thermal Infrared Radiometer) Japanese optical/infrared mechanically scanned radiometer, carried on *MOS-1* satellites. Wavebands: 0.50–0.70, 6.0–7.0, 10.5–11.5, 10.5–12.5 µm. Spatial resolution: 900 m (optical), 2.7 km (infrared). Swath width: 1500 km.

VTPR (Vertical Temperature Profile Radiometer)
1. Atmospheric temperature profiler carried by *NOAA-2* to *-5* satellites. Seven wavebands in the range 13.4 μm to 18.7 μm. Horizontal resolution 60 km.
2. See *SSE*.
3. See *SSH*.

VV-polarisation A term used in *radar*, to describe a signal that is both transmitted and received in *vertical polarisation*. This is a *co-polarised* mode of operation.

Walsh function See *Hadamard transform.*

Water, electromagnetic properties

Optical/infrared region

The diagram shows schematically the dielectric constant of pure water between 0.2 μm and 50 μm.

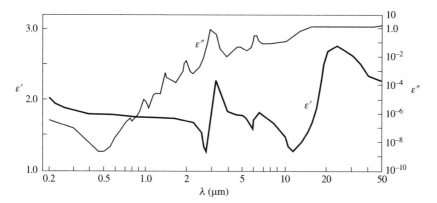

Typical attenuation coefficients (natural logarithms per metre) in the optical band:

λ (μm)	Bay	Coastal	Oceanic	Pure
0.4	0.9	0.4	0.15	0.07
0.45	0.5	0.2	0.09	0.05
0.5	0.4	0.2	0.09	0.05
0.55	0.3	0.2	0.12	0.05
0.6	0.4	0.3	0.2	0.18
0.65	0.5	0.4	0.4	0.3
0.7	0.7	0.6	0.6	0.6

Typical *emissivity* (10–12 μm): 0.99 (0.96 for distilled water)

Microwave region

The dielectric properties of pure water in the microwave region are well described by the *Debye equations* with temperature-dependent values of ε_p

and τ as follows. The value of ε_∞ can be taken as 5.0:

T (°C)	ε_p	τ (ps)
0	83.2	17.7
10	79.1	12.6
20	75.4	9.2
30	71.8	7.1
40	68.4	5.7

See also *sea water*.

Water vapour Atmospheric water vapour plays a major role in determining the Earth's climate, since it is the principal contributor to the greenhouse effect. Evaporation of water from the oceans, and its subsequent condensation to form *clouds*, is the major mechanism for heat transfer between the oceans and the atmosphere. For these reasons, knowledge of the distribution of atmospheric water vapour is fundamental to investigations of the global climate system. Measurements of water vapour can also provide a useful addition to numerical meteorological forecasting models.

With a knowledge of the temperature profile through the atmosphere (see *temperature sounding*), it is possible to use measurements of upwelling infrared radiation at the wavelengths of the appropriate absorption bands to determine the vertical distribution of water vapour. For example, on the *HIRS* instrument, three channels in the 6–8 μm water vapour band are employed and these provide data through most of the troposphere. Because a knowledge of the temperature profile is required in the inversion procedure and errors from the temperature retrieval are introduced, the humidity profiles obtained to date are not of a high quality. To obtain the humidity profiles, similar regression and physical inversion procedures to those described for *temperature sounding* are used. The introduction of the passive microwave *AMSU-A/MHR* instrument in the late 1990s should produce significantly more accurate water vapour sounding.

Integrated water vapour measurements are needed for meteorology and for atmospheric correction of other remote sensing observations, notable *radar altimeters*. These measurements can be obtained by integrating water vapour profiles obtained as described above, or by direct determination. Passive microwave radiometry (e.g. *MIMR, SSM/I, TMR*) or infrared radiometry (e.g. *GOES-Imager, MODIS, POLDER*) at water vapour absorption lines can be used to find the total water vapour content.

The principal infrared atmospheric absorption lines due to water vapour are at the following wavelengths: 0.72, 0.82, 0.93, 1.14, 1.38, 1.88, 2.7, 5.9, 6.3 μm. In the microwave band, there are important absorption lines at 22.2, 183, 325 and 390 GHz.

See also *atmospheric sounding*.

Wave height and wave spectra Knowledge of the *significant wave height* and energy spectrum of ocean waves is required for modelling the evolution of storms at sea and for studies of the energy and momentum exchanges between the atmosphere and the ocean. Ocean wave forecasts are important for shipping, for offshore exploration and for coastal protection.

Wave-height data can be obtained from *radar altimeter* measurements, and in future *laser profiler* instruments should enhance the available spatial resolution. Energy spectra can be obtained from imaging radar observations, particularly *synthetic aperture radar*.

Wavelength The distance between adjacent peaks or troughs, measured in the direction of propagation, in a harmonic wave. The wavelength λ and the *frequency f* are related by

$$v = f\lambda$$

where v is the *phase velocity* of the waves.

Wavenumber The wavenumber k of a wave is given by $2\pi/\lambda$, where λ is the *wavelength*. The wavevector **k** is a vector having magnitude k, directed in the propagation direction of the wave.

The wavenumber is sometimes specified as $1/\lambda$, rather than as $2\pi/\lambda$. In this case it is normally given the symbol σ.

W band Subdivision of the microwave region of the *electromagnetic spectrum*, covering the frequency range 56 to 100 GHz (wavelengths 3.0 to 5.4 mm).

WDVI (Weighted Difference Vegetation Index) See *vegetation index*.

Weather The weather is the varying atmospheric conditions which, when considered over an extended period of time, constitutes the *climate* of a region. The weather systems that bring about changes in atmospheric conditions vary across the Earth, but include mid-latitude depressions, anticyclones, fronts, hurricanes and tropical depressions. Satellite observations are now a vital element in meteorological studies, providing data for operational analysis and forecasting, and research investigations. Imagery allows the identification of many weather features, including mid-latitude depressions, fronts, areas of cloud located well away from the main weather systems, fog, tropical disturbances and thunderstorms. The broad scale upper flow can also sometimes be determined from cirrus cloud which is often present close to the jet streams. Temperature and humidity sounder observations provide one of the main forms of data for initialising the numerical weather prediction models in remote areas. Surface wind vectors over the ocean can be determined using scatterometer measurements and used to identify vortices and fronts. For research, satellite data are very valuable for investigating the three-dimensional structure of weather systems and their evolution in time. The most productive investigations are carried out using multi-sensor/multi-platform data sets.

See also *meteorology applications*.

Weighted Difference Vegetation Index See *vegetation index*.

Wetness See *tasselled-cap transformation*.

WFI (Wide Field Imager) Brazilian optical/near infrared *CCD* (pushbroom) imaging radiometer, planned for inclusion on *CBERS* and *SSR-1* satellites. Wavebands: CBERS: 0.63–0.69, 0.77–0.89 μm; SSR-1: 0.46–0.48, 0.63–0.69, 0.76–0.90, 1.23–1.25, 1.55–1.75 μm. Spatial resolution: CBERS: 260 m; SSR-1: 100–300 m. Swath width: CBERS: 890 km; SSR-1: 2200 km.

Whiskbroom scanner A common type of *scanning system* for optical, near infrared and thermal infrared radiation. Scanning in the direction perpendicular to the motion of the platform is achieved mechanically, using a rotating or oscillating mirror. Scanning in the perpendicular direction is achieved using the motion of the platform. Compare *pushbroom scanner* and *step-stare imager*.

Wide Field Imager See *WFI*.

Wide Field Sensor See *WiFS*.

Wien's law The wavelength at which the *spectral radiance* of *black-body radiation* is maximum is inversely proportional to the absolute temperature T of the radiation. This can be expressed as

$$\lambda_{\max} T = k,$$

where $k = 0.002\,99$ K m if the spectral radiance is L_λ, and $0.005\,10$ K m if it is L_ν.

WiFS (Wide Field Sensor) Indian optical/near infrared imaging radiometer, carried by *IRS-1C* and *IRS-1D* satellites. Wavebands: 0.62–0.68, 0.77–0.86 μm. Spatial resolution: 188 m. Swath width: 770 km.

WINDII (Wind Imaging Interferometer) Canadian–French instrument for measuring stratospheric wind speed and temperature, carried on *UARS*. WINDII is a nadir-viewing optical/infrared Michelson interferometer, used to measure Doppler shifts and temperature broadening of O_2 and OH emission lines. Height resolution: 20 km. Observation range: 80 to 300 km altitude. Accuracy: ± 10 m/s in wind speed.

URL: http://www-projet.cst.cnes.fr:8060/windii/Windii.html

Wind Imaging Interferometer See *WINDII*.

Wind lidar See *lidar*.

Windows, atmospheric Regions of the electromagnetic spectrum in which the atmosphere is reasonably transparent, so that remote sensing of the Earth's surface is a possibility. The principal atmospheric windows are listed below:

Optical/near infrared

Wavelength (μm)	Optical thickness	Source of variability
0.4–0.7	0.2–0.5	ozone, aerosols, clouds
2.0–2.5	0.1	water vapour, aerosols, clouds

Thermal infrared

Wavelength (μm)	Optical thickness	Source of variability
3.5–4.5	0.1–0.5	water vapour, clouds
8.0–9.0	0.1–0.7	water vapour, ozone, clouds
10–13	0.1–1.6	water vapour, clouds

Microwave

Frequency (GHz)	Optical thickness	Source of variability
0.1–20	0.01–0.1	water vapour, rain
25–40	0.1–0.2	water vapour, rain

Wind scatterometer See *scatterometer*.

Wind speed and velocity Consistent oceanic surface wind data of high quality and high temporal and spatial resolution are required to understand and predict the large-scale air-sea interactions which significantly influence both the atmosphere and the ocean. Such observations are needed to provide initial data and verification data for numerical weather prediction models, and to calculate surface fluxes of heat, moisture and momentum. Wind velocity data are also needed for such applications as numerical weather prediction, aircraft routeing, storm warning, and predicting the dispersal of marine pollutants.

Conventional surface wind velocity data over the oceans are provided by ships and buoys, but these data are extremely limited in coverage. As a result, analyses of surface-based wind observations often misrepresent atmospheric flow over large regions of the global oceans, and this contributes to the poor calculation of wind stress and sensible and latent heat fluxes in these regions. Data from satellite sensors provide the means to improve these analyses significantly by providing high resolution wind data over the global oceans.

In response to the wind blowing across it, the ocean surface responds on many wavelengths. This response provides a mechanism for the microwave remote sensing of ocean surface wind from space. The active sensing of the radar *backscatter* by centimetre-scale capillary waves allows the retrieval of ocean surface wind vectors with some ambiguity. The *Seasat* and *ERS scatterometers* and the NASA scatterometer *NSCAT* were all designed to take advantage of this phenomenon. Sea surface roughness measurements from *radar altimeters* can also yield wind speed data, though not direction.

Passive microwave remote sensing of the ocean surface also has the capability of retrieving ocean surface winds through the response of the microwave *emissivity* to surface roughness, and both the *SMMR* and *SSM/I* instruments have provided large datasets of ocean surface wind speed. Of these, the SSM/I has better accuracy, coverage and resolution.

SSM/I wind speeds have been combined with conventional surface observations and model-based fields in a variational analysis to generate an extensive SSM/I wind velocity data set. By March 1996, $8\frac{1}{2}$ years of SSM/I wind vectors had been processed since the operational phase of SSM/I began in July 1987. These data, which are accurate to typically $\pm 2\,\mathrm{m\,s}^{-1}$, are archived through the NASA Jet Propulsion Laboratory Distributed Archive Center.

Observations of wind velocity in the free atmosphere, above the Earth's planetary boundary layer, are required to represent the transports of atmospheric constituents and also for numerical weather prediction. Single-level wind velocity data can be inferred from *cloud* motions detected by *geostationary* satellite imagery (and, at high altitudes, from tracking *ozone* distribution). The technology for generating these wind data has evolved substantially. Originally, the movement of cloud elements was tracked manually. This was time-consuming, difficult, and often inaccurate, with the major problems being the assignment of the correct level to the wind and errors due to cloud dynamics. More objective procedures for tracking, based on correlation techniques, are now being used. Level assignment has improved somewhat but is still a significant source of error. Coverage is nearly global between latitudes $60°$ N and $60°$ S; however, sampling can still be a problem and this results in a low-speed bias for the jet streams. Cloud drift winds are most accurate in the tropics, with accuracies ranging from $3\,\mathrm{m\,s}^{-1}$ at low levels to 6 to $8\,\mathrm{m\,s}^{-1}$ at upper levels. Away from the tropical regions, high-altitude (geostrophic) winds can be inferred from atmospheric *temperature sounding* data.

Space-based *lidars* have been proposed to provide global wind velocity profile data. Experiments to simulate observing systems have indicated a very significant potential for lidar-generated wind data to improve atmospheric analyses and weather forecasting. At present both coherent and incoherent laser technologies are being pursued for this purpose, with the potential for launch early in the 21st century.

Within-class covariance See *canonical components*.

World reference system A coordinate system used to locate images from a satellite in an *exactly repeating orbit*. The system uses 'path' and 'row' numbers, the paths being parallel to the sub-satellite track and the rows being lines of constant latitude.

WVR (Water Vapour Radiometer) U.S. nadir-viewing passive microwave radiometer carried on *GFO*. Frequencies: 22 GHz, 37 GHz.

The WVR will be used to provide water-vapour corrections for the GFO *radar altimeter*.

X band Subdivision of the microwave region of the *electromagnetic spectrum*, covering the frequency range 5.2 to 10.9 GHz (wavelengths 28 to 58 mm).

X-SAR (X-band SAR) German *synthetic aperture radar* instrument carried on *SIR-C*. Frequency: X-band (9.6 GHz). Polarisation: VV. Incidence angle: 20° to 55°. Spatial resolution: 30 m. Swath width: 15 to 45 km.

Yellowness See *tasselled-cap transformation.*

Yellow substance Alternative name for **gelbstoff** (see *ocean colour*).

Zenith The direction vertically upwards, i.e. away from the Earth's centre. The opposite direction is the *nadir*.

Ziyuan-1 See *CBERS*.

Zonal harmonic See *dynamical form factor*.

Tables

Physical constants

speed of light *in vacuo*	c	$2.998 \times 10^8 \, \mathrm{m\,s^{-1}}$
Planck's constant	h	$6.626 \times 10^{-34} \, \mathrm{J\,s}$
elementary charge	e	$1.602 \times 10^{-19} \, \mathrm{C}$
rest-mass of electron	m_e	$9.109 \times 10^{-31} \, \mathrm{kg}$
permeability of free space	μ_0	$4\pi \times 10^{-7} \, \mathrm{H\,m^{-1}}$
permittivity of free space	ε_0	$8.854 \times 10^{-12} \, \mathrm{F\,m^{-1}}$
impedance of free space	Z_0	$376.7 \, \Omega$
Boltzmann's constant	k	$1.381 \times 10^{-23} \, \mathrm{J\,K^{-1}}$
gravitational constant	G	$6.673 \times 10^{-11} \, \mathrm{N\,m^2\,kg^{-2}}$
Stefan–Boltzmann constant	σ	$5.671 \times 10^{-8} \, \mathrm{W\,m^{-2}\,K^{-4}}$

Properties of the Earth

equatorial radius	a_e	$6.378 \times 10^6 \, \mathrm{m}$
polar radius	a_p	$6.357 \times 10^6 \, \mathrm{m}$
mass	M	$5.976 \times 10^{24} \, \mathrm{kg}$
dynamical form factor	J_2	1.083×10^{-3}
product of G with Earth's mass	GM	$3.986 \times 10^{14} \, \mathrm{m^3\,s^{-2}}$
surface gravitational field strength	g	$9.807 \, \mathrm{N\,kg^{-1}}$
atmospheric pressure at sea level	p_0	$1.013 \times 10^5 \, \mathrm{Pa}$
rotation period		$8.616 \times 10^4 \, \mathrm{s}$
orbital period		$3.156 \times 10^7 \, \mathrm{s}$

Bibliography

Avery, T. E. and Berlin, G. L. (1992). *Fundamentals of Remote Sensing and Airphoto Interpretation*. (Fifth edition.) New York, etc.: Macmillan Publishing Company.

Carsey, F. D. (editor) (1992). *Microwave Remote Sensing of Sea Ice*. (Geophysical Monograph no 68.) Washington DC: American Geophysical Union.

Colwell, R. N. (editor) (1983). *Manual of Remote Sensing*. (Second edition.) Falls Church, Virginia: American Society of Photogrammetry.

Cracknell, A. P. (1997). *The Advanced Very High Resolution Radiometer*. London: Taylor and Francis.

Curran, P. J. (1985). *Principles of Remote Sensing*. London, etc.: Longman.

Elachi, C. (1987). *Introduction to the Physics and Techniques of Remote Sensing*. New York, etc.: John Wiley and Sons.

Gurney, R. J., Foster, J. L. and Parkinson, C. L. (editors) (1993). *Atlas of Satellite Observations Related to Global Change*. Cambridge, etc.: Cambridge University Press.

Hord, R. M. (1982). *Digital Image Processing of Remotely Sensed Data*. New York, etc.: Academic Press Inc.

Ikeda, M. and Dobson, F. (1995). *Oceanographic Applications of Remote Sensing*. Boca Raton, Florida: CRC Press.

Jensen, J. R. (1996). *Introductory Digital Image Processing: a Remote Sensing Perspective*. (Second edition.) Englewood Cliffs, New Jersey: Prentice-Hall.

Kramer, H. J. (1996). *Observation of the Earth and its Environment*. (Third edition.) Berlin, etc.: Springer Verlag.

Leberl, F. W. (1990). *Radargrammetric Image Processing*. Norwood, MA: Artech House Inc.

Maling, D. H. (1973). *Coordinate Systems and Map Projections*. London: George Philip and Son.

Massom, R. (1991). *Satellite Remote Sensing of Polar Regions*. London: Belhaven Press.

Paterson, W. S. B. (1994). *The Physics of Glaciers*. (Third edition.) Oxford, etc.: Pergamon.

Rees, W. G. (1990). *Physical Principles of Remote Sensing*. Cambridge, etc.: Cambridge University Press.

Richards, J. A. (1993). *Remote Sensing Digital Image Analysis*. (Second edition.) Berlin, etc.: Springer Verlag.

Robinson, I. S. (1985). *Satellite Oceanography*. Chichester, etc.: John Wiley.

Sabins, F. F. (1978). *Remote Sensing: Principles and Interpretation*. San Francisco: W. H. Freeman and Co.

Schanda, E. (1986). *Physical Fundamentals of Remote Sensing.* Berlin, etc.: Springer Verlag.

Schowengerdt, R. A. (1997). *Remote Sensing: Models and Methods for Image Processing.* (Second edition.) New York, etc.: Academic Press Inc.

Schreier, G. (Ed.) (1993). *SAR Geocoding: Data and Systems.* Karlsruhe: Herbert Wichmann Verlag.

Scollar, I., Tabbagh, A., Hesse, A. and Herzog, I. (1990). *Archaeological Prospecting and Remote Sensing.* (1990). Cambridge etc.: Cambridge University Press.

Sonka, M., Hlavac, V. and Boyle, R. (1993). *Image Processing, Analysis and Machine Vision.* London, etc.: Chapman and Hall.

Stewart, R. H. (1985). *Methods of Satellite Oceanography.* Berkeley, etc.: University of California Press.

Swain, P. H. and Davis, S. M. (1978). *Remote Sensing: The Quantitative Approach.* New York, etc.: McGraw-Hill International Book Company.

Ulaby, F. T. and Elachi, C. (1990). *Radar Polarimetry for Geoscience Applications.* Dedham, MA: Artech House.

Ulaby, F. T., Moore, R. K. and Fung, A. K. (1981a). *Microwave Remote Sensing,* volume 1. Reading, MA: Addison-Wesley.

Ulaby, F. T., Moore, R. K. and Fung, A. K. (1981b). *Microwave Remote Sensing,* volume 2. Reading, MA: Addison-Wesley.

Ulaby, F. T., Moore, R. K. and Fung, A. K. (1986). *Microwave Remote Sensing,* volume 3. Dedham, MA: Artech House.

van de Hulst, H. C. (1957). *Light Scattering by Small Particles.* New York: John Wiley & Sons.